Girma Kibatu Berihie
Abaynesh Teka

Quality of Liver & Kidney Functional Tests among Medical Laboratories

Girma Kibatu Berihie
Abaynesh Teka

Quality of Liver & Kidney Functional Tests among Medical Laboratories

Western Amhara National Regional State of Ethiopia

LAP LAMBERT Academic Publishing

Impressum / Imprint
Bibliografische Information der Deutschen Nationalbibliothek: Die Deutsche Nationalbibliothek verzeichnet diese Publikation in der Deutschen Nationalbibliografie; detaillierte bibliografische Daten sind im Internet über http://dnb.d-nb.de abrufbar.

Bibliographic information published by the Deutsche Nationalbibliothek: The Deutsche Nationalbibliothek lists this publication in the Deutsche Nationalbibliografie; detailed bibliographic data are available in the Internet at http://dnb.d-nb.de.

Coverbild / Cover image: www.ingimage.com

Verlag / Publisher:
LAP LAMBERT Academic Publishing
ist ein Imprint der / is a trademark of
AV Akademikerverlag GmbH & Co. KG
Heinrich-Böcking-Str. 6-8, 66121 Saarbrücken, Deutschland / Germany
Email: info@lap-publishing.com

Herstellung: siehe letzte Seite /
Printed at: see last page
ISBN: 978-3-659-27219-6

Bahir Dar University
Department of Chemistry

Quality of Liver and Kidney Functional Tests among Public Medical Laboratories in Western Amhara National Regional State of Ethiopia

By Abaynesh Teka Nigussie (BSc)

Advisor: Girma Kibatu Berihie (PHD)

Department of Chemistry
Bahir Dar University
Bahir Dar, Ethiopia
June, 2011

Quality of Liver and Kidney Functional Tests among Public Medical Laboratories in Western Amhara National Regional State of Ethiopia

Thesis

Submitted in partial fulfillment of the requirements for the degree of

Master of Science in Chemistry

(Inorganic, Coordination and Bioinorganic Division)

By Abaynesh Teka Nigussie (BSc)

Advisor: Girma Kibatu Berihie (PHD)

Department of Chemistry
Bahir Dar University
Bahir Dar, Ethiopia

Thesis Approval Sheet

Quality of Liver and Kidney Functional Tests among Public Medical Laboratories in Western Amhara National Regional State of Ethiopia

By Abaynesh Teka Nigussie (BSc)

Approved by the Examining Board

______________________	______________
Chairman, Dep't Graduate Committee	Signature
______________________	______________
Advisor	Signature
______________________	______________
Internal Examiner	Signature
______________________	______________
External Examiner	Signature

Date: June 2011

Place: Bahir Dar

Signature

I, the undersigned, declare that the work reported herein represents my own ideas in my own words and wherever others' ideas or works have been included, I have adequately cited and referenced the original sources. I understand the non-adherence to the principles of academic honesty and integrity, misrepresentation/fabrication/falsification of any idea/data/fact/source will constitute sufficient ground for disciplinary action by the university and can also evoke penal action from the sources which have not been properly cited or acknowledged.

____

Signature

____

Name of the Student

____

Date and Place

Dedication

I want to dedicate this work for my families' especially to my father Teka Nigussie.

Table of Contents

Page

Table of Contents 6

Acknowledgement..7

List of Abbreviation..8

List of Figures..10

List of Graphs..11

List of Tables..12

Abstract..13

1. Introduction..14
2. Literature Review..16
   - 2.1 The Importance of Laboratory Quality..16
   - 2.2 Quality Management Systems..17
   - 2.3 Clinical Laboratory standards..22
     - 2.3.1 International Laboratory Standards..22
     - 2.3.2 Laboratory Standards in Ethiopia..24
   - 2.4 Quality Control and Quality Assessment..25
   - 2.5 External Quality Assessment in Clinical Chemistry Tests in ART Program..28
3. Statement of the Problem..34
4. Experimental Methods..35
   - 4.1 Study Area and Participant Laboratories in EQA Scheme..35
   - 4.2 Quality Control Materials..36
     - 4.2.1 Requirements of Control Materials..36
     - 4.2.2 Sample Preparation and Shipment for Chemistry Tests..36
     - 4.2.3 Measurements and Data Collecting Schemes..37
   - 4.3 Statistical Analysis..37
   - 4.4 Ethical Approval..38
5. Results and Discussion..39
   - 5.1 Qualitative Evaluation of Measurements..40
   - 5.2 Quantitative Evaluation of Measurements..41
6. Conclusions and Recommendations..59

References..61

Appendix..65

Acknowledgement

First of all I would like to express my deepest gratitude and special thanks to my supervisor Dr. rer. nat. Girma Kibatu Berihie for his constant and unreserved assistance in suggesting the title of the research, providing the necessary materials, giving valuable comments, proper guidance, critically editing the whole manuscript and facilitating all the necessary conditions starting from the beginning up to the end of this thesis. I would also like to express my heartfelt gratitude to my Co-advisor Mr. Yakob G/Michael, a clinical chemist in Regional Health Research Centre in Bahir Dar, for his consistent guidance and support, and in providing necessary materials for sample preparation and handling, Ato Mebet Admass, M.Sc.; a water civil engineer in Abay Tefasesie, for his support in mapping the study area using Geographical Information System.

My special thanks to Denekew Berihun, my husband, for sharing different up and downs during my study and travels to the different laboratories. His financial support, valuable ideas, moral support, and encouragement and facilitating all the necessary materials in the course of my study were tremendous. I am also very grateful to my youngest brother, Teshome Teka, for his support in collecting results from Metema (Shadie) Hospital. I would like to acknowledge my parents for their support in different ways, my sister, Bethlehem Teka, and my friends, Wubie Teshome and Yohannes Gidamu, for their invaluable support and advice.

My special thanks also go to all health personells in the Amhara National Regional State Health Office, Bahir Dar Regional Health Research Laboratory Center, the medical directors, laboratory coordinators and laboratory chemists working in hospitals or higher clinics with ART chemistry laboratories in North - West Amhara region who participated in this quality assessment study were cooperative in all standards. Finally I would like to thank Bahir Dar University for giving me the chance to learn in the masters program in chemistry, the chemistry department and to all staff members especially to Samuel Kasahun for their support in providing materials and supportive ideas.

List of Abbreviations

AIDS	Acquired Immuno Deficiency Syndrome
ALP	Alkaline Phosphatase
ALT	Alkaline Aminotransferase
ART	Antiretroviral Therapy
AST	Aspartate Aminotransferase
ATP	Adenine Triphosphate
BUN	Blood Urea Nitrogen
CLIA	Clinical Laboratory Improvement Amendment
CLSI	Clinical and Laboratory Standards Institute
CQI	Continuous Quality Improvement
EHNRI	Ethiopian Health and Nutrition Research Institute
EQM	External Quality Management
EQAS	External Quality Management Scheme
FDA	Food and Drug Administration
GCLP	Good Clinical Laboratory Practice
HIV	Human Immunodeficiency Virus
JCAHO	Joint Commission for Accreditation of Health care Organization
LIS	Laboratory Information System
OSHA	Occupational Safety and Health Administration
PPE	Personal Protective Equipment

QA Quality Assurance

QC Quality Control

QI Quality Improvement

QMS Quality Management Science

QP Quality Planning

SGOT Serum Glutamate Oxalate acetate Transaminases

SGPT Serum Glutamate Pyruvate Transaminases

USFDA United States Food and Drug Administration

List of Figures

Figure 2.1 Preanalytical, analytical and postanalytical phase of a laboratory work....................17

Figure 2.2 Path of work flow in a laboratory...18

Figure 2.3 Elements of quality system essentials..23

Figure2. 4 Types and rates of error in the three stages of the laboratory testing processes..........26

Figure 2.5 The structure of Alanine Aminotransferase (ALT)..30

Figure 4.1 Map of the Study Area...35

Figure 5.1 Scatter diagram for SGOT/AST. Plot of sample SGOT/AST (A) in **Hematrol N** values versus sample SGOT/AST (B) in **Hematrol P** values..41

Figure 5.2 Scatter diagram for SGPT/ALT. Plot of sample SGPT/ALT (A) in **Hematrol N** values versus sample SGPT/ALT (B) in **Hematrol P** values..43

Figure 5.3 Scatter diagram for ALP. Plot of sample ALP (A) in **Hematrol N** values versus sample ALP (B) in **Hematrol P** values...45

Figure 5.4 Scatter diagram for Creatinine. Plot of sample Creatinine (A) in **Hematrol N** versus sample Creatinine (B) in **Hematrol P** values..46

Figure 5.5 Scatter diagram for BUN. Plot of sample BUN (A) in **Hematrol N** values versus sample BUN (B) in **Hematrol P** values...48

Figure 5.6 Scatter diagram for Total Cholesterol. Plot of sample Total cholesterol A in **Hematrol N** values versus sample Total cholesterol B in **Hematrol P** values........................50

Figure 5.7 The average test results of GOT (A) and GOT(B) for each laboratory.....................51

Figure 5.8 The average GPT (A) and GPT (B) test results for each laboratory.......................52

Figure 5.9 The average ALP (A) and ALP (B) test results for each laboratory 53

Figure 5.10 The average BUN (A) and BUN (B) test results for each laboratory....................54
Figure 5.11 The average Creatinine (A) and Creatinine (B) test results for each laboratory 55
Figure 5.12 The average Total Cholesterol (A) and Total Cholesterol (B) test results for each laboratory..56

List of Tables

Table 5.1 Summary of the results of all test values in different laboratories.........................57

Table 5.2 Classification of all values according to their magnitude of error.........................58

Abstract

Medical laboratories are essential components of the health system and their major role is measurement of substances in body fluid for the purpose of diagnosis, treatment, prevention, and for greater understanding of the disease process. To achieve these aims the data generated has to be reliable for which strict quality control, quality management and quality assurance have to be maintained. The purpose of this study is to assess the performance of the west Amhara medical chemistry laboratories in Ethiopia in testing liver and kidney functions, which are usually used in monitoring HIV/AIDS, satisfactorily for this eight public medical laboratories and the Regional Health Research Laboratory Center found in west Amhara region of Ethiopia (working on ART) were participated in the study from February 18 up to March 15, 2011. Control samples of two serum pools, after a certain digestion, were distributed to these laboratories for estimation of liver and kidney functions tests. Test results reported from all participant laboratories were collected and statically evaluated against the control using SPSS 17; in addition, on-site observation and interview on the laboratories environment was performed to assess technical issues, safety procedures, the laboratory working area and related issues. About 64% of the 213 values reported fell outside of the allowable limits of errors for the chemistry tests and therefore were classified as unacceptable in many standards. A summary of the complete data is presented in the form of tables and scatter diagrams, it was found that there was a lack of accuracy and precision in many laboratories in west Amhara regions in Ethiopia so, the problems in developing country laboratories should be alleviated through collaborative approach among different stakeholders and through strengthening them in equipments and continuously training human resources. The use of quality management system and close follow-up with quality control and quality assessment schemes are recommended.

1. Introduction

Laboratory service is an essential component of the health care and laboratory diagnostic support for the investigation of epidemics and surveillance of endemic diseases cannot be successful without adequate and organized laboratory facilities and trained human resource[1]. Reliable clinical laboratory services are essential for diagnostic, control and treatment of disease in general. One area where the laboratory plays a great role in disease diagnostic, monitoring and control is its role in ART programs for HIV/AIDS patients.

Laboratory support is critical in all the areas of HIV diagnosis and management[2] and data supplied by the clinical chemistry laboratory play an important role in clinical decision making. Measures to control the quality of the results in an HIV diagnostic laboratory are extremely important, because the consequences of either false positive or false negative results are huge[3]. As such, it is essential that the data provided be reliable. According to a study made by the US Institute of Medicine medical errors contributed to more than 1 million injuries in the United States annually[4]; at least a portion of these may be attributed to errors in data provided by the clinical chemistry laboratories. Everyday, the clinical chemistry laboratory is faced with many opportunities for errors that may be hazardous to patients as well as to persons involved in laboratory work. Thus laboratories should regularly assess their activities and control their qualities.

Quality laboratory reports help the physician to establish proper diagnosis rapidly and support better health care for patients. Apart from providing accurate results, maintaining quality in laboratory services also help in generating confidence among patients in the health care system, creating a good reputation for the laboratory, reducing costs by avoiding repetition of tests, sustains motivation in laboratory staff, helps in the accreditation of the laboratory and prevents legal suits and associated complications.

The accuracy and precision of clinical chemistry results are frequently assessed in developed countries for different tests[5]. However, quality control schemes are not common in many developing countries. In the case of our country, no research has been done on external quality assessment especially in the Amhara National Regional State. The ANRS Health Bureau has made surveys on clinical laboratory data, number of HIV patients, ART services, number of patients who follows ART, etc in terms of norms representing individual performance and technical matters than on real needs and performance of the laboratories as a whole. It is important to know that clinical laboratories has a better reliability if assessed in terms of selected test results than other data that do not clearly show the quality of the services.

HIV/AIDS is one of the greatest health crises ever faced by humanity. This pandemic has already killed 25 million people since its discovery in 1980s[6]. Today, more than 40 million people are living with HIV and each year three million die of HIV/AIDS. However, most of these deaths could be prevented if they had access to antiretroviral therapy (ART) using antiretroviral drugs. The ART strategy, however, demands competent laboratory services parallel to the use of antiretroviral drugs by patients and their doctors.

The government of Ethiopia has been undertaking different interventions to curb HIV/AIDS destructions on the societies; among which ART program is the one[7]. Quality of the ART service is a comprehensive and challenging process; need multiple sources of supports and supervision from clients, providers, managers, families, communities and others. Specially, the laboratory service is one strong arm of the health care providers that helps to monitor and control the disease in the health system. The laboratory service should thus be standardized, properly managed and regularly assessed for quality.

It is now a common understanding that getting reliable or accurate laboratory results is crucial for following and deciding the health status of a patient. The current work focuses on assessing the accuracy and precision of common chemistry tests (renal and liver functional tests, usually used for monitoring HIV/AIDS patients on ART programs) in medical laboratories found in north - western Amhara region of Ethiopia. Quality assessment studies were made among eight district ART service provider laboratories and the Regional Health Research Laboratory Centre. The district laboratories include: government hospital chemistry laboratories in Bahir Dar, Debretabor, Gondar, Metema (Shadie), Debark, Funeteselam, Debremarkos and Mota. This study assesses the quality of these medical chemistry laboratories in testing liver and kidney functions that are usually used in monitoring HIV/AIDS patients on ART.

2. Literature Review

2.1 The Importance of Laboratory Quality

Clinical laboratories are the key partners in patient safety and the medical laboratory service is one of the major factors which directly affect the quality of health care.

Laboratories produce test results that are widely used in clinical and public health settings, and health outcomes depend on the accuracy of the testing and reporting. Laboratory results are known to influence 70% of medical diagnoses[8]. If inaccurate results are provided, the consequences can be very significant:

- unnecessary treatment; treatment complications
- failure to provide the proper treatment
- delay in correct diagnosis and,
- additional and unnecessary diagnostic testing.

These consequences result in increased cost in time, personnel effort, and often in poor patient outcomes.

In order to achieve the highest level of accuracy and reliability, it is essential to perform all processes and procedures in the laboratory in the best possible way. The laboratory is a complex system, involving many steps of activity and many people. The complexity of the system requires that many processes and procedures be performed properly. Thus, laboratory quality management system is very important for achieving good laboratory performance[9].

Laboratory quality can be defined as accuracy, reliability, and timeliness of the reported test results. The laboratory results must be as accurate as possible, all aspects of the laboratory operations must be reliable, and reporting must be timely in order to be useful in a clinical or public health setting.

Studies in different developing countries show that services in ART laboratories are not up to the standards which may decline the safety and effectiveness of the drugs, may have negative impact on adherence, satisfaction and continuity of care and treatment; besides it hampers expansion and accessibility of the services[10]. For example, a cross-sectional descriptive study conducted to assess the quality standards of health facilities providing antiretroviral treatment (ART) in Dares Salaam from May to July in 2005 indicated that there were inadequate trained personnel, laboratory

equipments, laboratory environment, antiretroviral drugs, weak quality control and quality management system[11]. There were inadequate confidential places for counseling; and information system was found to be weak. Another study also indicated that the effect of poor quality laboratory prevents eligible patients who need to start ART and comprehensive HIV care and treatment from using all the designated facilities[12].

2.2 Quality Management Systems

A quality management system can be defined as coordinated activities to direct and control an organization with regard to quality. In a quality management system, all aspects of the laboratory operation, including the organizational structure, processes, and procedures, need to be addressed to assure quality[13].

There are many procedures and processes that are performed in the laboratory and each of these must be carried out correctly in order to assure accuracy and reliability of testing. An error in any part of the cycle can produce a poor laboratory result[14]. A method of detecting errors at each phase of testing is needed if quality is to be assured.

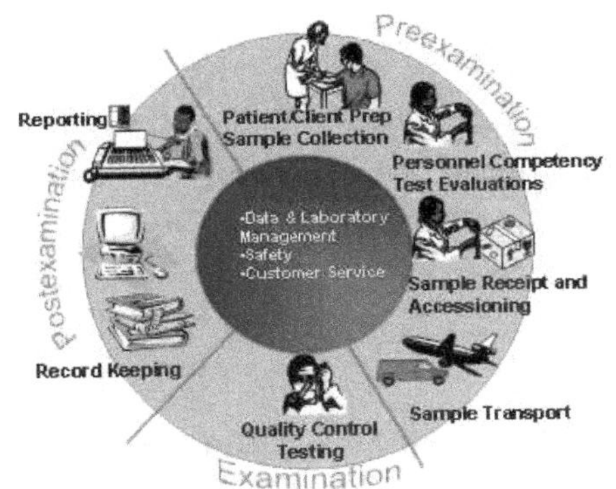

Figure 2.1 preanalytical, analytical and postanalytical phase of a laboratory work

The entire set of operations that occur in testing is called the *Path of Workflow*. The Path of Workflow begins with the patient and ends in reporting and results interpretation.

The concept of the Path of Workflow is a key to the quality model or the quality management system, and must be considered when developing quality practices. For example, a sample that is damaged or altered as a result of improper collection or transport cannot provide a reliable result. A medical report that is delayed or lost, or poorly written, can negate all the effort of performing the test well.

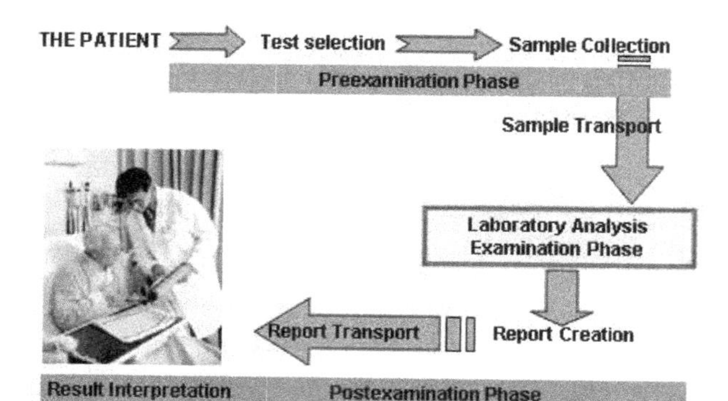

Figure 2.2 Path of work flow in a laboratory.

The complexity of the laboratory system requires that many factors must be addressed to assure quality in the laboratory. Some of these factors include:

- the laboratory environment
- quality control procedures
- communications
- record-keeping
- competent and knowledgeable staff
- good quality reagents and equipment.

The Development of Good Laboratory Practice

The understanding on the need of quality laboratories in the health service resulted in the formation of Good Laboratory Practice (GLP) by the US Food and Drug Administration, continued by the Organization for Economic Co-operation and Development in1980[15].

The basic role of the clinical chemistry laboratory according to GLP is to provide accurate qualitative and/or quantitative data on biological specimens and it focuses on technical performance of the laboratory technicians especially on the analytical phase, it targets on people as a source of problem and quality is measured only by the analytical methods and procedures[16]. Nevertheless, the analysis and assays performed by the laboratories are subject to variance due to different things beyond the personnel performance[17]. The following are some of the sources of variance in results which contribute to errors, and the possible means to minimize them.

Collection and transport of specimens

Incorrect procedures for collection of specimen are the source of variance which account for far more errors than any other source. Collection of specimens may start with preparation of patients, e.g. fasting and postprandial blood and urine samples. Further steps include avoidance of haemolysis of blood samples: using appropriate anticoagulants/preservative: correct identification and labeling of specimens: and prompt and proper transport of specimens to the laboratory.

Specimen handling in the Laboratory

Improper handling of specimens which reach the laboratory is another source of variance. For result to be accurate and reliable certain precautions need to be taken, depending on the nature of specimens. These include: prevention of evaporation from specimen tubes. Prevention of exposure to light, correct serum separation procedures, storage conditions etc.

Analytical methods and procedures

Sources of variance at this stage include;

- Reagents which are standard or have deteriorated.
- Incorrect amounts of reagents and/or specimens at any stage of procedure (pipetting error)
- Improper mixing of specimens and reagents.
- Failure to maintain correct temperatures or timings.
- Faulty apparatus due to improper/inadequate maintenance calibration or even erratic power supply.
- Errors in reading and calculation of results.

Competency of personnel

All the personnel involved in laboratory work have completed their respective courses of study/ training and have certificates of qualification. Even so, for certain analytical methods a lot of experience is required for achieving consistent and reliable results, especially where some steps of procedure have to be done manually. Hence this becomes a source of variance. New staff should therefore be given adequate on-job training by experienced workers. A full time supervisor e.g. Pathologist is advisable to deal promptly with technical problems that can arise.

Environment

A proper environment or working condition is cooperative to good quality work, hence getting good quality results. Laboratories with poor working conditions/ environment would have more difficulties in maintaining the quality of work performed by the laboratory staff, making it a source of variance. For achieving a good environment the laboratory should have adequate space, proper lay out efficiency, clean, well ventilated with adequate lighting and comfortable ambient temperature.

Work load

A very high work load of staff ratio could adversely affect the quality of work performed. On the other hand too low a ratio could induce laxity/inattention of staffs and is equally unproductive. To maintain the quality of work, the work load of staff ratio should therefore be optimal.

From the above discussion on different sources of variance on test results it can be seen that there are many factors which can contribute to variability and errors in results or data provided by the clinical chemistry laboratories. Fortunately, not all variations amount to errors which are clinically significant since there is an acceptable range of values for most parameters analyzed. When diagnosis and prognosis of some disease condition or their response to therapeutic regimens depend on laboratory data, significant errors may be hazardous to the patient, and a quality laboratory is a must to minimize the hazard. To minimize or even avoid hazards arising from clinical chemistry laboratories, the staff should strive for " the right result at the right time on the right specimen from the right patient " constant vigilance is the key and quality assurance which encompasses all the factors described above is mandatory[18].

In 1989 Laffel and Blumenthal[19], have noted that the traditional quality assurance approaches (GLP) in health care have several limitations in terms of the definition of quality, satisfying the need of the customers and evaluating the performance of the organization as a whole. In the same year, they introduce the use of modern quality management science which has been successful in other industrial sectors into the health care organizations. Modern Quality Management System (QMS) offers new opportunities for improving the quality of health care.

Framework for Modern Quality Managing System

Modern industrial QMS broadens the definition of customers so that physicians and nurses are considered a laboratory's immediate customers, even though the patient is the ultimate customer, and emphasizes the continuous improvement of quality and targets processes rather than people as the source of problems.[20] People are considered to be an organization's primary resource and are to be continually developed through training and education, provides participatory mechanisms (project teams, quality circles, suggestion programs) to involve everyone in problem solving and quality improvement. In this way, quality becomes a pervasive issue that extends beyond technical performance to organizational effectiveness in the delivery of service. Quality becomes the culture of the organization and is reflected by its entire people and all their actions. Berwick[21] explains the need for QMS by describing the traditional management philosophy as the "theory of bad apples." In this approach, managers collect and analyze data and blame problems on the people who are doing the work. The strategy for improving quality according to the modern QMS is to weed out the bad apples. According to Berwick, however, health care organizations must abandon this approach and focus instead on the continuous improvement of quality.

In addition to this, health care organizations should work to design their delivery systems so as to provide the services they need. They must recognize that their customers outside the laboratory are many; they include patients, physicians, nurses, administrators, other departments, external regulate agencies, and payers. According to O'Connor, [22] tomers judge the quality of the services not by reliability alone, but also by responsiveness, competence, access, courtesy, communication, credibility, security, understanding consumer needs, and other tangibles such as facilities, equipment, and the appearance of personnel. Therefore, quality services depend on providing a "totality of features and characteristics that conform to the stated or implied needs of these users or customers". To do jobs well, one must understand its customers' needs and expectations and establish standard quality goals that guide the design, evaluation, implementation, and operation of the daily work processes.

2.3 Clinical Laboratory Standards

Laboratory standardization for integrated diagnosis of diseases is necessary for all levels of the laboratory system to define the services required at each facility. This can be done through infrastructure upgrades, trainings, quality assurance, equipment maintenance, supply chain initiatives, and other strategies to assist the clinical laboratories and upgrade their ability to provide laboratory services for integrated diseases[24].

2.3.1 International Laboratory Standards

The ISO 9000 documents provide guidance for quality in manufacturing and service industries, and can be broadly applied to many other kinds of organizations[25]. ISO 9001:2000 addresses general quality management system requirements and apply to laboratories. There are two ISO standards that are specific to laboratories:

1 ISO 15189:2007. Medical laboratories–Particular requirements for quality and competence. Geneva: International Organization for Standardization.

2 ISO/IEC 17025:2005. General requirements for the competence of testing and calibration laboratories. Geneva: International Organization for Standardization.

The International Organization for Standardization (ISO) 15189 (medical laboratories: particular requirements for quality and competence) is the current internationally recognized standard for medical laboratory practice.[26] This standard provides a framework for a laboratory to plan and operate a medical testing laboratory with an effective quality management system (QMS) that has strong elements of quality assurance, quality control, and quality improvement.

It is not only focuses on assay performance, but also has a holistic approach for global medical patient care, targeting all processes from preanalytical to postanalytical procedures. It further encompasses personnel laboratory safety and medical laboratory ethics. An award of ISO 15189 accreditation is a formal international recognition that a laboratory performs medical laboratory tests that are reliable, accurate, and reproducible and that the results are reported within an acceptable time frame[27].

Another important international standards organization for laboratories is the Clinical and Laboratory Standards Institute, or CLSI, formerly known as the National Committee for Clinical Laboratory

Standards (NCCLS). CLSI has two documents that are very important in the clinical laboratory.

1 CLSI/NCCLS. *Application of a Quality Management System Model for Laboratory Services.*

2 CLSI/NCCLS. *A Quality Management System Model for Health Care.*

CLSI provides the necessary background information and infrastructure to develop a quality management system that will meet healthcare quality objectives and be consistent with the quality objectives of each organization or service. This guideline provides a structure for a comprehensive, systematic approach to build quality into the healthcare organizations, assess the organization and implement quality improvements. This document, used with the relevant discipline-specific companion document for individual service areas, can provide the means to apply this model to their respective operations. CLSI uses a consensus process involving many stakeholders for developing standards. CLSI developed the quality management system model; this model is based on twelve quality system essentials (QSE), and is fully compatible with ISO laboratory standards.

The Modern Quality Management System Model

When all of the laboratory procedures and processes are organized into an understandable and workable structure, the opportunity to ensure that all are appropriately managed is increased. The modern quality management system model for medical laboratories recently developed by CLSI involves all activities in the processes, and is also fully compatible with ISO standards.[23] The quality model organizes all of the laboratory activities into *twelve quality system essentials.* These quality system essentials are a set of coordinated activities that serve as building blocks for quality management. Each must be addressed if overall laboratory quality improvement is to be achieved. Assuring accuracy and reliability throughout the Path of Workflow depends on good management of all of the quality essentials.

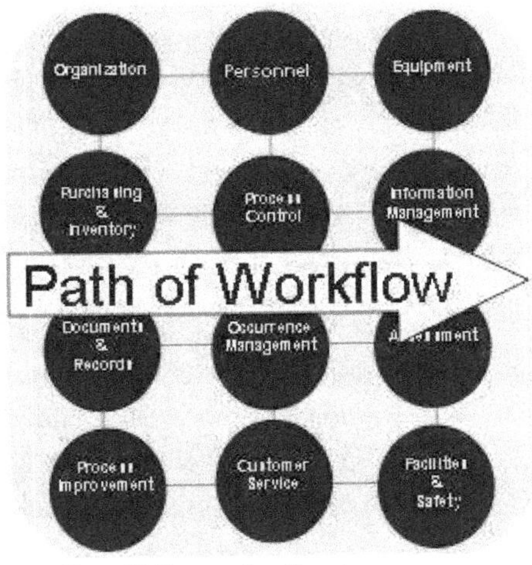

Figure 2.3 Elements of quality system essentials

2.3.2 Laboratory Standards in Ethiopia

In Ethiopia, the Italians were the first to establish health laboratories during the Second World War in 1928 E.C. In more than 50 years of medical laboratory existence in Ethiopia there has not been a national standard for medical laboratories besides targets and protocols set by the WHO which is usually not simple to implement in developing countries. The Ethiopian Health and Nutrition Research Institute, EHNRI, which is a result of the merger in April 1995 E.C. of the former National Research Institute of Health, the Ethiopian Nutritional Institute and the Department of Traditional Medicine of the Ministry of Health is the strong arm in the Ethiopian Ministry of Health which works with programs to make laboratory services in public facilities efficient and standardized as to the targets set by WHO. Ethiopian Health and Nutrition Research Institute developed a plan for growth and transformation of the country on laboratory standards.

The Ethiopian Health and Nutrition Research Institute in its strategic plan (2010 – 2015 G.C) have pledged to support laboratories through capacity building, quality assurance programs, infrastructure development, training and maintenance[28]. EHNRI will also design to conduct regular monitoring and evaluation of laboratory services in an effort to improve services and meet acceptable laboratory standards. As a result, EHNRI aims to develop an affordable and sustainable system, whereby quality laboratory services are accessible to all Ethiopians while also providing reliable and high-quality results to guide and support clinical decision making throughout the country. Different guidelines, manuals, SOPs and formats have to be developed to standardize the laboratory system and standards will be set such that all critical health issues can be addressed by Ethiopia's laboratory system.

Standardization and building capacity at the regional and federal laboratories will enhance their abilities and quality in performing specialized and referral tests, and implement Regional External Quality Assessment Scheme. All laboratories will be included in external quality assessment schemes at the national and regional level through improving laboratory standards and training them with different quality assurance systems such as external quality assessment schemes which increase the confidence of health care practitioners to use laboratory data.

2. 4. Quality Control and Quality Assessment

2.4.1 Quality Control

A major role of the clinical laboratory is the measurement of substances in body fluids or tissues for the purpose of diagnosis, treatment or prevention of disease, and for greater understanding of the disease process. To fulfill these aims the data generated has to be reliable for which strict quality control has to be maintained. Quality control is defined as the study of those sources of variation, which are the responsibility of the laboratory, and the procedures used to recognize and minimize them.

Quality control mainly focuses on analytical stages of experiments, and it involves consideration of a reliable analytical method. Reliability of the selected method of a clinical laboratory test is determined by four indicators. Two of these, accuracy and precision, reflect how well the test method performs day to day in a laboratory. The other two, sensitivity and specificity, deal with how well the test is able to distinguish disease from absence of disease.

The accuracy and precision of each test method are established and are frequently monitored by the professional laboratory personnel. Sensitivity and specificity data are determined by research studies and are generally found in medical literature. Although each test has its own performance measures and appropriate uses, laboratory tests are designed to be as precise, accurate, specific, and sensitive as possible. These basic concepts are the cornerstones of reliability of test results and provide the confidence health care providers have in using the clinical laboratory.

Studies of laboratory errors have documented that a higher percentage of errors occur in the pre analytic and post analytic processes than in analytic processes. The figures often quoted on average are 45% for errors in preanalytic processes, 10% for analytic errors, and 45% for post analytic errors based on a US based study done in 1988[29] before the implementation of the current Clinical Laboratory Improvement Amendments (CLIA) regulations. As a consequence of this expected distribution of errors, laboratories are urged to focus their attention on preanalytic and post analytic processes to improve patient safety[30]. The final CLIA rule reflects this emphasis on increased quality assessment for preanalytic and post analytic processes and proposes a reduction in quality control (QC) for analytic processes[31].

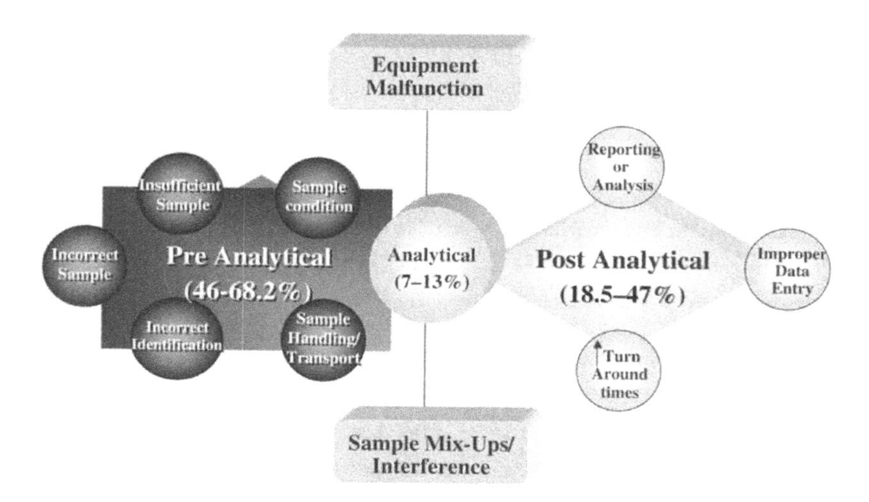

Figure 2.4 Types and rates of error in the three stages of a laboratory testing process. (From: Errors in clinical laboratories or errors in laboratory medicine. Mario Plebani. Clin Chem Lab Med 2006; 44 page 752.)

2.4.2 Quality Assessment

In a broad sense, a medical laboratory can be regarded as a factory producing results of tests on clinical samples taken from patients at the request of medical staff. Clearly, the quality of the laboratory product is critical to the treatment of the patient. The most important element in the quality of the test result is its *accuracy* (or correctness). If the laboratory result is falsely negative, there is a chance that the patient's illness will go undiagnosed, or, in some cases, will be incorrectly diagnosed. Possible consequences for the patient include continued suffering, or even death. If the patient happens to be suffering from an infectious disease, there is a risk of continued transmission to the patient's family and close contacts. If the laboratory result is falsely positive, there will be an incorrect diagnosis and the patient is likely to receive unnecessary treatment, such as hospitalization and therapy with toxic drugs.

Accuracy is understandably the element of quality that is given most attention. However, in addition to accuracy, which can be measured by various means, there is a need to consider many other aspects of the laboratory's operation. These include:

- Is the laboratory environment appropriate for the work being performed?
- Are the staff numbers adequate for the workload?
- Are the operating procedures up-to-date and followed by all staff

- Are all staff adequately trained in the test processes?
- Are the results produced economically?
- Is the laboratory working in collaborative partnership with its clients, the medical staff?

So, QA is a complete system of creating and following procedures and policies to aim for providing the most reliable patient laboratory results and to minimize errors in the preanalytical, analytical and postanalytical phases. QA also includes analyzing known samples called quality control, QC samples along with unknown (patient) samples to test for analytical problems. When QC samples do not produce accurate and precise results, it can be assumed that any patient results obtained at the same time are also erroneous. Following a set of guidelines for acceptance or rejection of patient results based on the QC results helps to assure reliability of the analysis. Specific rules for assuring quality regarding the patient specimen through collection, analysis, and reporting are also important aspects of QA. Thus, preanalytical, analytical, and postanalytical variables need to be considered and minimized in order to have valid test results. Assessment actually leads to assuring quality by a close follow up of laboratories when actions taken to correct problems become permanent changes in policies, procedures, and behaviors.

In relation to quality assessment, a study made in Ethiopia to know the status of HIV screening laboratories in different parts of the country by Belete Tegbaru et. al.[33] indicated that 64% of the laboratories had no close follow up and supervision which affects the performances of the laboratories.

Laboratory's overall performance should be monitored through a series of regular activities and laboratories should be close together and should be a close follow up by responsible organizations. Quality assessment activities produce various pieces of information: some will be quantitative (e.g. rate of errors in a test panel), and some will be qualitative (e.g. poor maintenance of equipments). The combined results serve as a rational basis on which the performance of the laboratory can be assessed. QA programs can be internal or external.

1 Internal quality assessment (IQA) is the set of procedures undertaken by the staff of a laboratory for the continuous monitoring of operations and results, in order to decide whether the results are reliable enough to be released.

2 External quality assessment (EQA) involves inter-laboratory surveys by independent EQA organizers. This study also focuses on external quality assessment.

2.5. External Quality Assessment in Clinical Chemistry Tests for ART Program

It is essential that laboratories are engaged in quality assurance programmes and participate successfully in inter-laboratory comparison programmes (i.e. External Quality Assessment Schemes (EQAS) also known as Proficiency Testing Schemes) [34]. Participation within a recognized external quality assurance (EQA) scheme benefits include gaining information on the relative performance of different methods and knowledge, about one's own ability to perform tests, report results accurately, as well as gaining the confidence of clinicians and patients.

ART laboratories participating on EQA schemes can be tested for different parameters depending on the interest of the external quality assurance body. The different tests for monitoring the health status of HIV patients are: *viral load, CD4 count, complete blood count* and *chemistry screen or chem panel tests*[35]. These tests are blood tests and are the most comprehensive tests available to monitor the health of individuals living with HIV depending on patient's health and their response to treatment regimen, especially on anti-HIV drugs which may result problems on liver and kidney functions. Most doctors run these tests every three to six months for their patients. Since these tests are used to monitor an overall health through comparisons of tests over time, it is important that patients get their lab work done when they are first diagnosed or when they start their treatment regimen to provide baseline information for future comparisons.

Chemistry Screen or Chem Panel Tests

These tests are general screening tests that are used to measure major body organs (heart, liver, kidneys, pancreas, etc), muscles and bones functions by measuring specific chemicals in the blood. These tests are essential in the detection of infections or side effects from medications[36]. Chemistry panel tests should be done at least every six months, and more often in people who are experiencing symptoms or taking drugs that can adversely affect blood values. Important chemistry tests in medical laboratories include: Electrolytes, Liver Enzyme Tests, and Kidney Function Tests.

A/ Electrolytes

Electrolytes are positively or negatively charged molecules (ions) that play important roles in cellular activity and heart and nerve function. Normally electrolyte levels are regulated by the kidneys, and any excess is excreted in the urine[37]. Most healthy people can get all the electrolytes and other minerals they need by eating a balanced diet. Electrolytes include sodium, potassium, calcium, chloride and bicarbonates. Electrolyte imbalances in the body may signal malnutrition, kidney problems, or dehydration (which may be caused by prolonged vomiting or diarrhea).

B/ Liver Enzyme Tests: Alanine Aminotransferase (ALT), Aspartate Aminotransferase (AST) and Alkaline Phosphatase (ALP)

The liver is the largest and one of the most important organs in the body. As the body's "chemical factory" it regulates the levels of most of the biomolecules found in the blood, and acts with kidneys to clear the blood of drugs and toxic substances. The liver metabolizes toxic substances, alters their chemical structure, makes them water soluble, and excretes them in bile. Laboratory tests for total protein, albumin, ammonia, enzymes and cholesterol are markers for the function of liver. Tests for cholesterol, bilirubin, ALP, and bile salts are measures of the excretory function of the liver. The enzymes ALT, AST, GGT, LD are tested as markers for liver injury, if the liver has been damaged or its function impaired. Problems in the liver can damage liver cells and release their enzymes into the blood, which can be measured as an indicator of liver cell damage. Deviations of these markers from reference values due to liver injury or disease tell the physician that something is wrong with the liver. Having a healthy liver is important to everybody, but it is particularly important for people with HIV as the liver plays a key role in breaking down and processing medicines used to treat HIV and other infections and in addition some anti-HIV drugs can cause side-effects that affect the liver and if somebody is taking them, doctors will want to see if his or her liver is suffering any ill-effects because of them.

Alanine aminotransferase (ALT)

Alanine aminotransferase (ALT) also called Serum Glutamic Pyruvate Transaminase (SGPT) is an enzyme present in hepatocytes (liver cells). When a cell is damaged, it leaks this enzyme into the blood stream, where it is measured. ALT rises dramatically in acute liver damage, such as viral hepatitis or from medication like paracetamol (acetaminophen) overdose and anti-HIV drugs.

In biochemistry, aminotransferase is an enzyme that catalyzes a type of reaction between an amino acid and α-keto acid. To be specific, this reaction (transamination) involves removing the amino group from the amino acid, leaving behind α-keto acid, and transferring it to the reactant α-keto acid and converting it into an amino acid. The enzyme ALT transfers an amino group from the amino acid alanine to a ketoacid acceptor (oxaloacetate) which is important for glycolysis (degradation of glucose

$C_6H_{12}O_6$ in to pyruvate $CH_3COCOO^- + H^+$), and for Kerbs citric acid cycle (a metabolic pathway involved in the chemical conversion of carbohydrates, fats and proteins in to ATP, carbon dioxide and water).

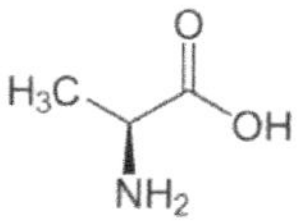

Alanine

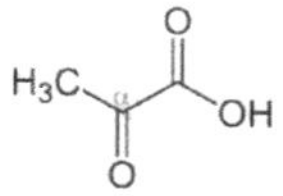

Pyruvic acid

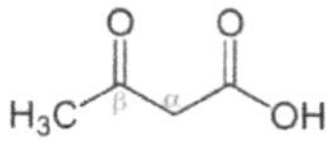

Acetoacetic acid

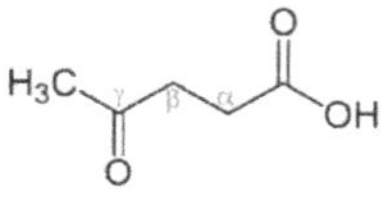

Levulinic acid

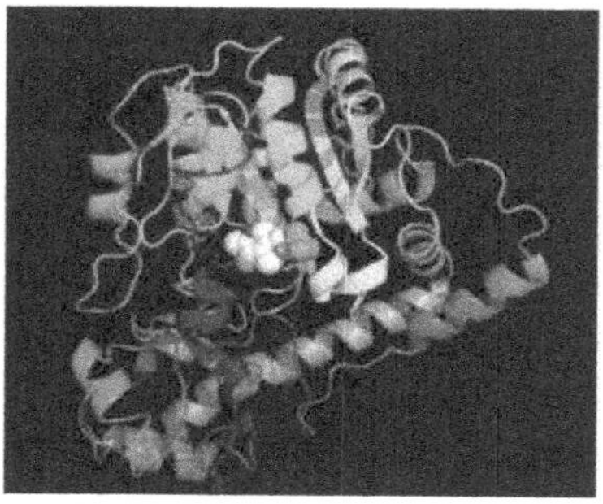

Alanine aminotransferase

Figure 2.5 The structure of Alanine aminotransferase

The enzyme is very sensitive to necrotic or inflammatory liver injury. Consequently, if ALT or direct bilirubin are increased, then some form of liver disease is likely. If both are normal, then liver disease is unlikely.

Aspartate aminotransferase (AST)

Aspartate aminotransferase (AST) also called Serum Glutamic Oxaloacetic Transaminase (SGOT) is similar to ALT in that it is another enzyme associated with liver parenchymal cells. It is raised in acute liver damage, but is also present in red blood cells and cardiac and skeletal muscle and is therefore not specific to the liver. The ratio of AST to ALT is sometimes useful in differentiating between causes of liver damage.[38] Elevated AST levels are not specific for liver damage, and AST has also been used as a cardiac marker.

Alkaline phosphatase, ALP

ALP is a hydrolase enzyme responsible for removing phosphate groups from many types of molecules, including nucleotides, proteins, and alkaloids. The process of removing the phosphate group is called *dephosphorylation.* As the name suggests, alkaline phosphatases are most effective in an alkaline environment. It is sometimes used synonymously as basic phosphatase.[39] Alkaline phosphatase (ALP) is an enzyme in the cells lining the biliary ducts of the liver. ALP levels in plasma will rise with large bile duct obstruction, intrahepatic cholestasis or infiltrative diseases of the liver. ALP is also present in bone and placental tissue, so it is higher in growing children (as their bones are being remodeled).

In humans, alkaline phosphatase is present in all tissues throughout the entire body, but is particularly concentrated in liver, bile duct, kidney, bone, and the placenta. Humans and most other mammals contain the following alkaline phosphatase isozymes:

- ALPI – intestinal
- ALPL – tissue non-specific (liver/bone/kidney)
- ALPP – placental (Regan isozyme)

High ALP levels can show that the bile ducts are blocked.

The levels of these enzymes in the blood can vary considerably; the normal range of ALT is between 5 and 60 IU/L (international units per liter), the normal range of AST is between 5 and 43 IU/L and the normal range of ALP is between 20 and 70 IU/L[40]. The levels of these enzymes are increased if there is liver damage. Common causes include damage from alcohol, hepatitis, medications or other drugs.

Total Cholesterol Test

It is a fatty substance that circulates in the blood; cholesterol is an important component of cell membranes, certain hormones, vitamin D, and bile acids. A healthy total cholesterol level is 120-200 mg/dL[41]. Elevated total cholesterol (hypercholesterolemia) is known to increase the risk of cardiovascular disease, but it is more useful to look at specific types of cholesterol. Low-density lipoproteins (LDL) so-called "bad" cholesterol can deposit cholesterol in artery walls, causing atherosclerosis (hardening of the arteries). But high-density lipoproteins (HDL) so-called "good" cholesterol helps clear cholesterol from the body and may reduce the risk of cardiovascular disease. In some studies, higher HDL levels have been associated with more robust viral load reductions in people taking anti-HIV therapy and abnormal levels of fats may be caused by long term HIV infection as well as some anti-HIV medications, specifically protease inhibitors[42].

The National Cholesterol Education Program recommends that people try to achieve a total cholesterol level below 200 mg/dL, an LDL level below 100 mg/dL, and an HDL level of at least 40 mg/dL[43].

C/ Kidney Tests: **BUN and Creatinine**

The main job of the kidneys is to filter out waste products. They reabsorb what is needed and remove the waste in urine. The most important waste products are excess sodium and water. Each kidney contains about a million filtering units called nephrons. They:

- eliminate wastes from the body,
- regulate the volume and pressure of blood, and
- control levels of electrolytes and blood acidity.

HIV can cause kidney failure due to HIV infection of kidney cells. This is known as HIV-Associated Nephropathy or HIVAN. Kidney disease is one cause of morbidity and mortality in HIV-infected patients. However, the rate of kidney disease in patients with HIV has gone down significantly since the introduction of modern antiretroviral therapy (ART.) and about 30% of people with HIV may have kidney disease. If kidney disease advances, it can cause heart disease.

Blood Urea Nitrogen (BUN) is a normal metabolic waste product that is excreted by the kidney, it is a byproduct of protein breakdown. It is a measure of the amount of nitrogen in the blood in the form of urea.

Creatinine

The most commonly used test to measure kidney function is Creatinine. Creatinine is a waste product of protein digestion and muscle breakdown. High levels of creatinine indicate a problem in the kidneys' ability to do their job of removing waste from the body.

It is the metabolic bi-product of creatine, is an organic acid that assists the body in producing muscle contractions. Creatinine is found in blood stream and in muscle tissue and removed from the blood by the kidneys as urine.

Both **BUN** and **Creatinine** always appear on a chem. screen report, and are important blood values associated with kidney health. Normal BUN levels should be between 8 and 23 milligrams per deciliter of blood (mg/dL); normal creatinine levels should be between 0.7 and 1.3 mg/dL[44]. These tests are very important to watch by people taking drugs that may affect the kidneys, such as Foscavir (foscarnet) and Vistide (cidofovir) [45].

In general, laboratory chemistry results for electrolytes, liver enzyme tests and kidney function tests play a great role in monitoring the health status of HIV patients on ART drugs. The objectives of this study were: to assess the performance of the north - west Amhara medical laboratories in measuring the above indicator tests used in ART program services and to see how much their laboratory results have closeness with control ranges and target values and the reproducibility of the results within individual laboratories.

3. Statement of the Problem

The government of Ethiopia has been undertaking different interventions to curb HIV/AIDS destruction on communities; among which the ART service program is one and the laboratory service is one strong arm for the health care providers for the HIV patients in the ART program. So, the laboratory Service of the west Amhara medical laboratories should be standardized, properly managed and regularly assessed for quality in addition to serving the patients with antiretroviral drugs.

Objective of the Study

3.1 General Objectives

The objective of this study is to assess the performance of the north - west Amhara medical chemistry laboratories in Ethiopia in testing liver and kidney functions, which are usually used in monitoring HIV/AIDS, satisfactorily. The study is to see how much individual laboratory results have closeness with control ranges and target values and to see if the results from Regional Health Research Laboratory Centre and results within individual laboratories are reproducible. The purpose of quality assessment is to reinforce improvement of programs.

3.2 Specific Objectives

- To determine the performance of the north - western Amhara Medical laboratories in delivery of services to the satisfaction of customers, the real needs and performance of the organization as a whole.
- To evaluate the performances and contributions of laboratories in monitoring and controlling HIV disease.
- To compare and contrast the results obtained from the control, the Regional Health Research Laboratory Centre and the eight district hospital medical laboratories in north - west Amhara region.
- To reinforce improvement programs on quality management systems in diagnostic laboratories.

4. Experimental Methods

4.1 Study Area and Participant Laboratories in EQA Scheme

External quality assessment was conducted among hospital based medical laboratories in west Amhara region of Ethiopia from February to March, 2011. The region covers a landmass of 98,235.36 sq. km, having an estimated 10,826,171 people with a density of 110 Person/Km^2. From this area, eight public medical chemistry laboratories in government hospitals and the Bahir Dar Regional Health Research Center which gives laboratory services for ART service users participated in the study.

Figure 4.1 Map of the Study area

The nine clinical chemistry laboratories included in the study are Lab. A (Debark hospital) for north

Gonder, Lab. B (Debremarkos hospital) for Gozamin Woreda, Lab. C (Debretabor hospital) for Farta Woreda, Lab. D (Bahir Dar Feleghiwot hospital) for Bahir Dar Zuria, Lab. E (Funeteselam hospital) for Jabi Tehnan Woreda, Lab. F (Gonder hospital) for Gonder Zuria, Lab. G (Shedie hospital) for Metema, Lab. H (Mota hospital) for Hulet Ej Enese Woreda and Lab. I (the Bahir Dar Regional Health Research Center).

4.2 Quality Control Materials

All chemicals and reagents used in this work were analytical grade reagents and were used with no further purification. A pooled serum control samples **Hematrol N (Human GmbH, Max-Planck-Ring 21 – D-65205 Wiesbaden, CS-HNS-N/022, INF 1351102, 04-2008-1)** and **Hematrol P (Human GmbH, Max-Planck-Ring 21 – D-65205 Wiesbaden, CS-HPS-P020, INF 1351202, 09-2008-2)** were used in this work. Distilled water was used to work with the control samples.

The control samples (**Hematrol N** and **Hematrol P**) were obtained from Bahir Dar Regional Health Research Laboratory Centre, which were stored in a refrigerator with a temperature 2^0C- 8^0C until sample digestion done.

4.2.1 Requirements of Control Materials

The control samples should have known amounts of measurable indicators of Liver functional tests and kidney functional tests on ART patients. Control samples have known stated values, normal range values and method described for analysis for the analyte of interest. These values can be used as a reference to compare the results of each laboratory with those values listed for the controls.

4.2.2 Sample Preparation and Shipment for Chemistry Tests

Initially 5 ml of distilled water were added to six bottles of the control samples, **Hematrol N** and **Hematrol P** (3 bottles each), then the solutions in the bottles were shacked manually for 10 minutes to increase the solubility. After that the three bottles of **Hematrol N** solutions were collected in one beaker (labeled A) and the three bottles of **Hematrol P** solutions were collected in another beaker (labeled B). Besides, 54 necked bottles (27 for beaker A and 27 for beaker B) were prepared to contain control solutions from each beaker. The necked bottles are labeled as:

A_1, A_2, A_3... and A_{27} and B_1, B_2, B_3 ... and B_{27}

A specimen solution of 0.5 ml from beaker A was then aliquot in to each necked bottle A_1, A_2, A_3

$\ldots A_{27}$, and similarly 0.5 ml of the specimen solution from beaker B was aliquot to each necked bottle $B_1, B_2, B_3 \ldots B_{27}$.

After reconstitution, six necked bottle specimen aliquots (6 specimens x 9 Laboratories = 54) were then distributed to each laboratory for analysis; the sample bottles were numbered differently in each case. The bottles had no other markings and were packed in an ice box for safe storage and transport by keeping it in suitable condition.

4.2.3 Measurement and Data Collecting Schemes

Participants were asked to analyze six specimens ($3A_n$ and $3B_n$) for SGOT/AST, SGPT/ALT, ALP, BUN, Creatinine and Total Cholesterol. The results were collected with a report sheet (Appendix B), and the values obtained with information concerning the methods and the type of instrument used.

4.3 Statistical Analysis

The allowable limits of error for each test were calculated by means of an empirical formula. This formula is based on the premise that errors should not exceed one quarter of the normal range. This formula is as follows:

$$\text{Allowable limits of error (in \%)} = \pm\left(\frac{\frac{1}{4}\text{of the normal range}}{\text{Mean of the normal range}}\right)\times 100\%$$

Since the normal range of SGOT/AST is taken to be 27.5 - 43.9 IU/L, then the allowable limits of error calculated by this formula were ± 10 %. The maximum limits for any determination, however, were set at ± 10 %, even though in some cases the values calculated by the above formula exceeded this figure because of the premise that errors should not exceed 10 %. For this study, then the allowable limits of error have been established as follows: for SGPT/ALT = ± 10 %, for ALP = ± 10 %, for BUN = ± 10 %, for Creatinine = ± 10 %, and for total Cholesterol = ± 7 %. Both sample A_n and sample B_n had the same allowable limits of errors.

An evaluation of the accuracy of each component was determined by calculating the percentage of results which fell outside of these allowable limits of errors. To evaluate precision, standard deviations and scattered diagrams were used.

4.4 Ethical Approval

The research proposal "***Quality of Liver and Kidney Functional Tests among Public Medical Laboratories in Western Amhara National Regional State of Ethiopia***." was reviewed by the Amhara National Regional State Health Office and was found to be ethically clear in terms of its protocols and potential risks for the general public. In addition to this the medical directors and laboratory technologists of the eight hospitals and the Regional Health Research Laboratory Center had accepted the idea and show consent to cooperate and facilitate the work by every possible means.

5. Results and Discussion

5.1 Qualitative Evaluation from on-site Observation

A cross-sectional study using interviews and detailed on-site observation/supervision to assess technical issues, safety procedures, laboratory management and related issues revealed the following:

- Since HIV is a blood born disease, the ART program needs a great attention and safety for both the patient and the laboratory staff, but many of the laboratory workers did not give attention for the work, even they did not wear a lab coat, they did tests with out a glove, they did not put contaminated things on their suitable places, they did not give an attention for test result reading and interpretation from computers.
- The environment in which laboratory testing performed must be conductive to efficient operations that do not compromise the safety of the staff or the quality of the preanalytic, analytic and post analytic processes. Laboratory work area must have sufficient space so that there is no hindrance to the work or employee safety, all floors, walls, and bench tops of the laboratory must be clean and well maintained. But some of the surveyed laboratories did not have adequate space. Sample collection, testing and data interpretation takes place on the same room and these results a confusion of identifying once blood sample with the other, exposure of the laboratory worker and other staffs with different infections.
- According to GCLP, laboratories must employ an adequate number of qualified personnel to perform all of the functions associated with the volume and complexity of tasks and testing performed within the laboratory, but almost all of the supervised laboratories did not have adequate personnel, a single individual collects and tests the blood samples of ART following patients.
- To maintain the quality of the laboratory work, the work load of the staff should be optimal and every laboratory personnel should get appropriate payment for the work. But in few laboratories the researcher got individuals who were not happy by their work and told that the work load and their salary did not match and this results in not giving due attention for the work, their customers and were not easy to communicate.
- Standard operating procedures (SOPs) are critical for maintaining consistent test performance.

The laboratory must write SOPs for all laboratory activities to ensure the consistency, quality, and integrity of the generated data. Current SOPs must be readily available in the work areas and accessible to testing personnel[29]. In the surveyed laboratories all laboratories have SOP but technicians did not seem to follow it either because it was not readily available or they depended more on their past experience.

5.2 Quantitative Evaluation of Measurements

All the eight medical laboratories and the Regional Health Research Laboratory Centre, which were invited to participate in this scheme completed the study with an exception of the laboratory in Metema (Shadie) which was not able to do all the tests asked (e.g. Total cholesterol, ALP Creatinine and BUN) due to improper temperature conditions, except the Regional Health Research Laboratory Center and Gonder laboratory, Total cholesterol was not tested by others due to lack of reagent. The results obtained were summarized in scatter diagrams and in table.

Evaluation of Individual Chemistry Tests

The following graphs are scatter diagrams for the samples tested by the different clinical chemistry laboratories by plotting sample A, **Hematrol N**, values against their corresponding sample B, **Hematrol P**, values. These diagrams show not only the actual values obtained but also the correlation between the sets of values. The vertical and horizontal lines are drawn to outline the acceptable ranges (based on the stated value). Each point represents a different laboratory test result.

SGOT/AST Measurements

The normal range of SGOT/AST (A) in **Hematrol N** is 27.5 - 43.9 IU/L, the allowable limits of error calculated was ± 10 %. The stated and mean values for SGOT/AST (A) were 35.7 IU/L and 35.5 IU/L respectively. Therefore, the acceptable ranges of the values which would be considered as acceptable test results based on stated and mean values were from 31.9 - 38.9 IU/L and from 32 - 39 IU/L, respectively.

For SGOT/AST (B) in **Hematrol P**, the normal range is 117 – 187 IU/L, the allowable limits of error Calculated is ± 10 %. The stated and mean values for SGOT/AST (B) were 152 IU/L, and 152.2 IU/L, respectively; therefore, the acceptable ranges of values which are considered as acceptable test results based on stated and mean values are from 136.8 – 167.2, IU/L and 137 – 167.4, IU/L respectively.

From the reported values for the test SGOT/AST (A), 18.5 % of the reported values fell below the acceptable range and also 18.5 % fell above the acceptable range with a total error of 37 % fell out of the acceptable range of values and for SGOT/AST (B), 59 % of the test results fell below the acceptable range and 15 % of the test results fell above the acceptable range, totally 74 % of the results fell out of the acceptable range based on the stated value.

The Graph of SGOT/AST (A) in **Hematrol N** versus SGOT/AST (B) in **Hematrol P** values based on the stated value is as follows:

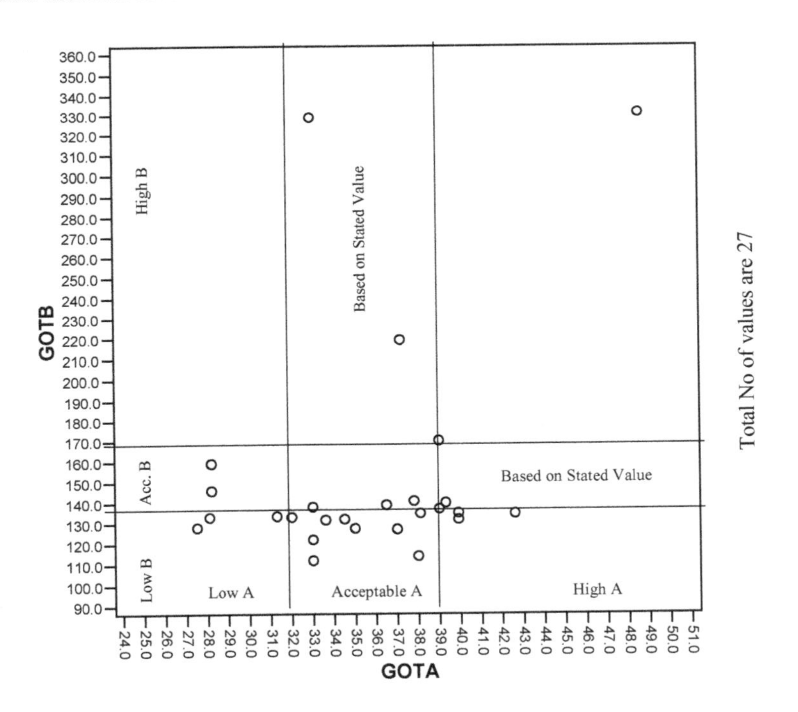

Figure 5.1 Scatter diagram for SGOT/AST. Plot of sample SGOT/AST (A) in **Hematrol N** values versus sample SGOT/AST (B) in **Hematrol P** values.

The centre section contains points plotted from values which were both acceptable for SGOT/AST (A) and SGOT/AST (B), and the upper right hand section, points plotted from values which were both too high (that is, too high for sample A and for sample B) and the lower left points plotted from values which were both too low (that is, too low for sample A and too low for sample B)

If errors had been only due to lack of accuracy, say improper standardization resulting in values A and B both being either to high or too low, then all the points not in the central areas would have fallen either in to the upper right (high A, High B) or lower left (low A, low B) sections.

From the surveyed laboratories Lab. A, Lab. C, Lab. D and Lab. I SGOT/AST (A) results 100 % fell in the acceptable range and Lab. F 67 % of the results fell in the acceptable range and Lab. B, Lab. E, Lab. G and Lab. H only 33 % of the results fell inside the acceptable range. So, Lab. A, Lab. C, Lab. D, and Lab. I had good performance on testing SGOT/AST (A) and for SGOT/AST (B) Lab. A had a better performance even than Lab. I based on the stated value.

SGPT/ALT Measurements

Since the normal range of SGPT/ALT (A) in **Hematrol N** is taken to be 24.8 – 39.6 IU/L, then the allowable limits of error calculated was ± 10 %. The stated and mean values of SGPT/ALT were 32.2 IU/L and 30.8 IU/L, respectively; therefore, the acceptable ranges of values which are considered as acceptable test results based on stated and mean values are from 29 – 35.4 IU/L, and 27.8 – 33.8 IU/L, respectively.

For SGPT/ALT (B) in **Hematrol P** the normal range is 112 – 180 IU/L, then the allowable limits of error is ± 10 %. The stated and mean values were 146 IU/L and 143 IU/L, respectively; therefore, the acceptable ranges of values which are considered as acceptable test results based on stated and mean values are from 131.4 – 160.6, IU/L and 128.7 – 157.3, IU/L respectively.

From the reported values for the test SGPT/ALT (A), 33 % of the reported values fell below the acceptable range and also 15 % fell above the acceptable range with a total error of 48 % fell out of the acceptable range of values and for SGPT/ALT (B), 74 % of the test results fell below the acceptable range and 11 % of the test results fell above the acceptable range, totally 85 % of the results fell out of the acceptable range.

The Graph of SGPT/ALT (A) in **Hematrol N** versus SGPT/ALT (B) in **Hematrol P** values based on the stated value is as follows:

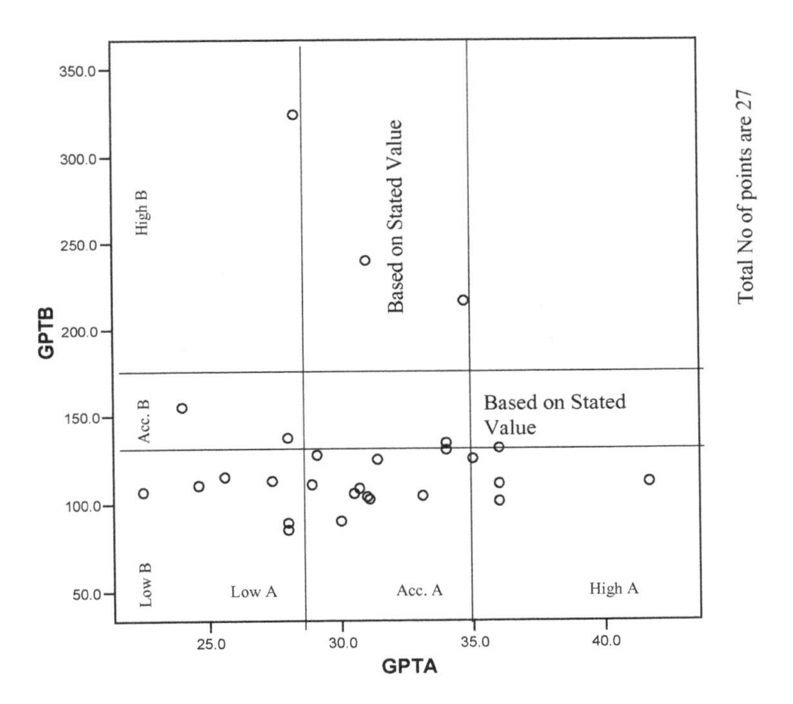

Figure 5.2 Scatter diagram for SGPT/ALT. Plot of sample SGPT/ALT (A) in **Hematrol N** values versus sample SGPT/ALT (B) in **Hematrol P** values.

From the surveyed laboratories only Lab.E SGPT/ALT (A) test results 100 % fell in the acceptable range and Lab. F and Lab. H all of the results fell out of the acceptable range. So, Lab. E had good performance on testing SGPT/ALT (A) and Lab. F and Lab. H show the worst performance on testing SGPT/ALT. For SGPT/ALT (B) only Lab. C has a better performance. The other laboratories including Lab. I had poor performance on testing SGPT/ALT (B).

ALP Measurements

The normal range of ALP (A) in **Hematrol N** is taken to be 164 – 273 IU/L, then the allowable limits of error calculated was ± 10 %. The stated and mean values of ALP (A) were 218 IU/L and 220 IU/L, respectively; therefore, the acceptable ranges of values which were considered as acceptable test results based on stated and mean values were from 196.2 – 239.8 IU/L and 198 - 242 IU/L respectively.

For ALP (B) in **Hematrol P** the normal range was 395 – 658 IU/L, then the allowable limits of error is ± 10 %. The stated and mean value were 526 IU/L and 510 IU/L, respectively, therefore, the acceptable ranges of values which are considered as acceptable test results based on stated and mean values were from 473.4 – 578.6 IU/L and 459 – 561 IU/L, respectively.

From the reported values for the test ALP (A), 50 % of the reported values fell below the acceptable range and also 25 % fell above the acceptable range with a total error of 75 % fell out of the acceptable range of values and for ALP (B), also 50 % of the test results fell below the acceptable range and 25 % of the test results fell above the acceptable range, totally 75 % of the results fell out of the acceptable range.

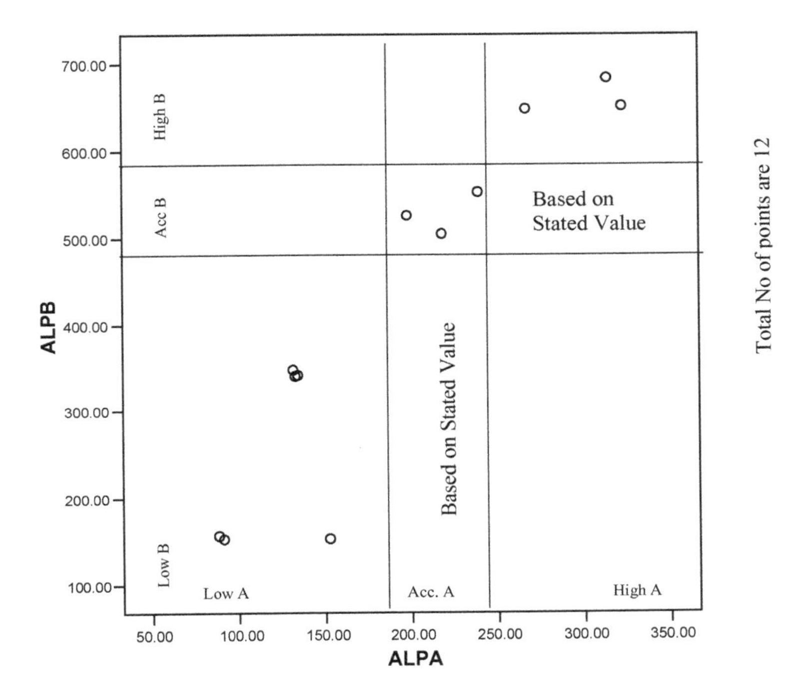

Figure 5. 3 Scatter diagram for ALP. Plot of sample ALP (A) in **Hematrol N** values versus sample ALP (B) in **Hematrol P** values.

The result shows that only Lab. D, Lab. F, Lab. H and Lab. I did ALP test, the rest did not due to lack of reagent. Lab. I ALP (A) test results all fell in the acceptable range and Lab. D, Lab. F and Lab. H all of the results fell out of the acceptable range. So, Lab. I had good performance on testing ALP (A). For ALP (B) also Lab. I had a better performance. The other laboratories had poor performance on testing ALP (B).

Creatinine Measurements

The normal range of Creatinine (A) in **Hematrol N** is taken to be 1.31 – 2.05 mg/dL, then the allowable limit of error calculated was ± 10 %. The stated and mean values of Creatinine (A) were 1.68 mg/dL and 1.6 mg/dL, respectively; therefore, the acceptable ranges of values which were considered as acceptable test results based on stated and mean values were from 1.5 – 1.85 mg/dL and 1.44 – 1.76, mg/dL respectively.

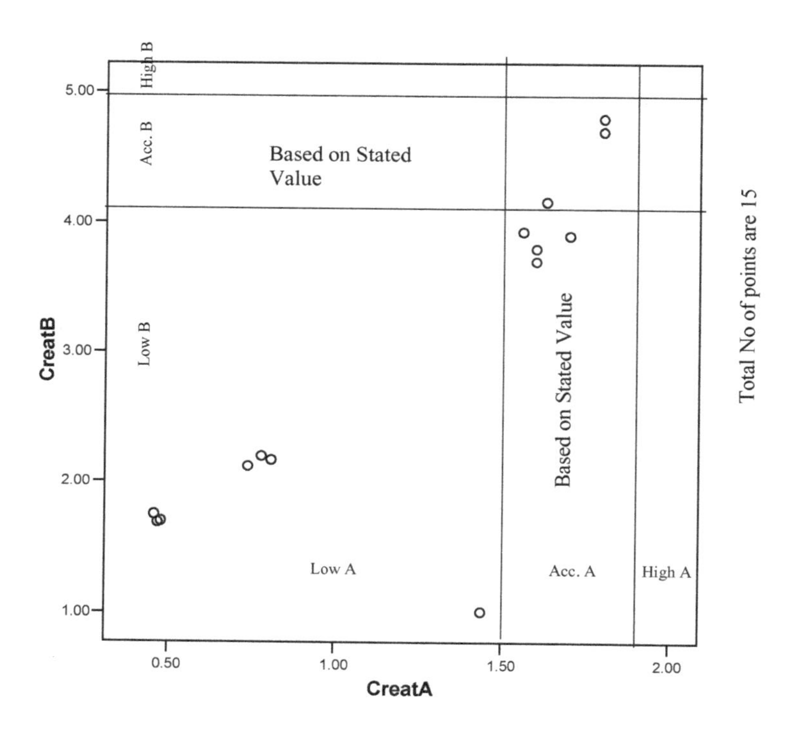

Figure 5.4 Scatter diagram for Creatinine. Plot of sample Creatinine (A) in **Hematrol N** versus sample Creatinine (B) in **Hematrol P** values.

For Creatinine (B) in **Hematrol P** the normal range was 3.55 – 5.55 mg/dL, then the allowable limits of error is ± 10 %. The stated and mean value were 4.55 mg/dL and 4.8 mg/dL, respectively, therefore, the acceptable ranges of values which were considered as acceptable test results based on stated and mean values were from 4.1 – 5.0 mg/dL and 4.3 – 5. 3 mg/dL, respectively.

From the reported values for the test Creatinine (A), 53 % of the reported values fell below the acceptable range and no values fell above the acceptable range and a total error of 53 % fell out of the acceptable range of values and for Creatinine (B), totally 80 % of the test results all fell below the acceptable range

The result shows that only Lab. A, Lab. B, Lab. C, Lab. E, and Lab. H did Creatinine test, the rest did not due to lack of reagent. Lab. A, Lab. C and Lab. H Creatinine (A) test results all fell in the acceptable range and Lab. B, and Lab. E all of the results fell out of the acceptable range. So, Lab A, Lab. C and Lab. H had good performance on testing Creatinine (A). For Creatinine (B) only Lab. A has a better performance. The other laboratories had poor performance on testing Creatinine (B).

BUN Measurements

The normal range of BUN (A) in **Hematrol N** was taken to be 50.4 – 78.8 mg/dL, then the allowable limits of error calculated was ± 10 %. The stated and mean values of BUN (A) were 64.6 mg/dL and 62 mg/dL respectively; therefore, the acceptable ranges of values which were considered as acceptable test results based on stated and mean values were from 58.1 – 71 mg/dL and 55.8 – 68.2, mg/dL respectively.

For BUN (B) in **Hematrol P** the normal range was 115 – 179 mg/dL, then the allowable limits of error is ± 10 %. The stated and mean value were 147 mg/dL and 136.1 mg/dL, respectively, therefore, the acceptable ranges of values which are considered as acceptable test results based on stated and mean values were from 131.4 – 160.6 mg/dL and 122.5 – 149.7, mg/dL respectively.

From the reported values for the test BUN (A), 57 % of the reported values fell below the acceptable range and also 4.8 % fell above the acceptable range with a total error of 61.8 fell out of the acceptable range of values and for BUN (B) 23.8 % of the test results fell below the acceptable range and 47.6 % of the test results fell above the acceptable range, totally 71.4 % of the results fell out of the acceptable range.

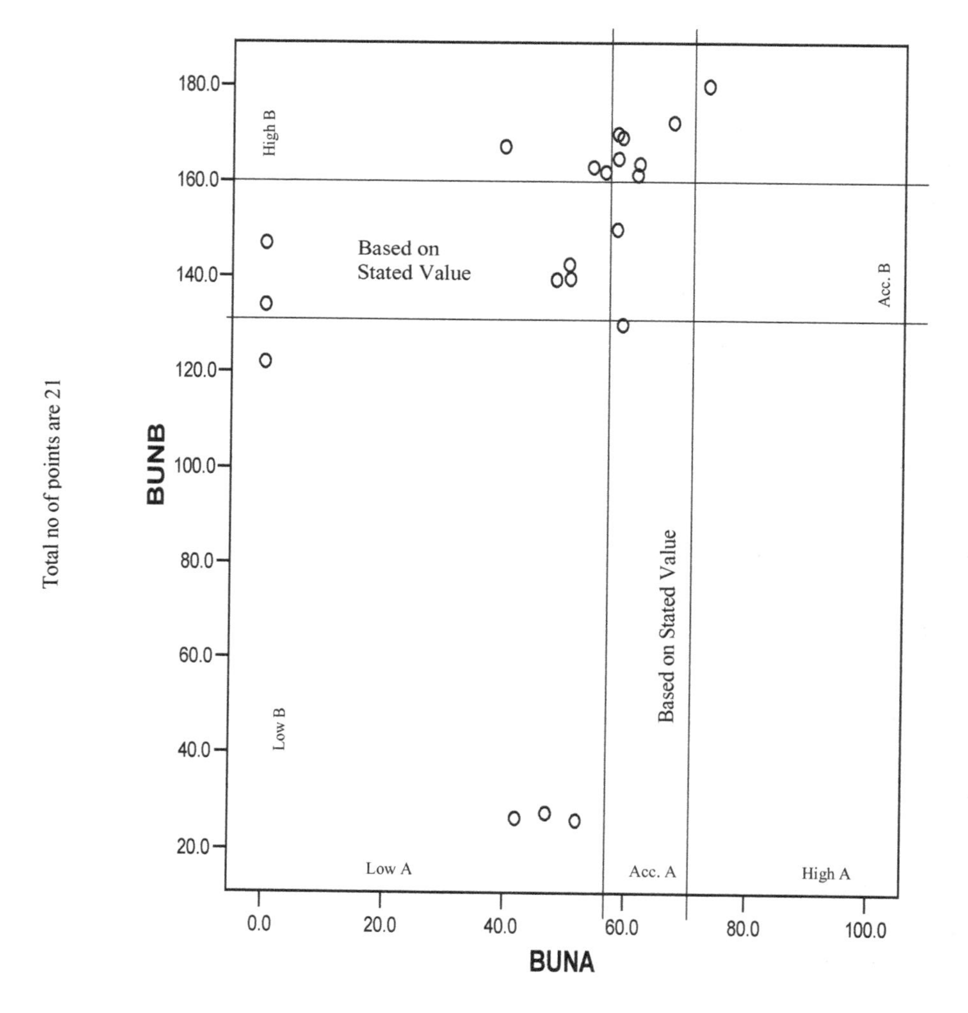

Figure 5.5 Scatter diagram for BUN. Plot of sample BUN (A) in **Hematrol N** values versus sample BUN (B) in **Hematrol P** values.

The result shows that Lab. F and, Lab. G, did not do Creatinine test due to lack of reagent. And from the rest laboratories, only Lab. H and Lab. I had good performance on testing BUN (A). For BUN (B) only Lab. A, Lad. D and Lab. I had a better performance. The other laboratories had poor performance on BUN (B).

Total Cholesterol Measurements

The normal range of Total cholesterol (A) in **Hematrol N** is taken to be145 – 193 mg/dL, then the allowable limit of error calculated was ± 7 %. The stated and mean values of Total cholesterol (A) were 169 mg/dL and 164 mg/dL respectively; therefore, the acceptable ranges of values which are considered as acceptable test results based on stated and mean values are from 157.2 – 180.8 mg/dL and 147.6– 180.4 mg/dL respectively.

For Total cholesterol (B) in **Hematrol P** the normal range was 217 – 287 mg/dL, then the allowable limits of error is ± 7 %. The stated and mean value were 252 mg/dL and 244 mg/dL respectively, therefore, the acceptable ranges of values which were considered as acceptable test results based on stated and mean values are` from 234.4 – 259.6 mg/dL and 227 – 261 mg/dL respectively.

From the reported values for the test Total Cholesterol (A), 50 % of the reported values fell below the acceptable range and also no value left above the acceptable range and for Total Cholesterol (B), 33.3 % of the test results fell below the acceptable range and 33.3 % of the test results fell above the acceptable range, totally 66.6 % of the results fell out of the acceptable range.

Only two laboratories Lab. F and, Lab. G, did Total cholesterol test the rest did not do this test due to lack of reagent. Lab. F had a better performance on testing Total Cholesterol (A) than Lab. I. For Total Cholesterol (B) both of them did not have a good performance.

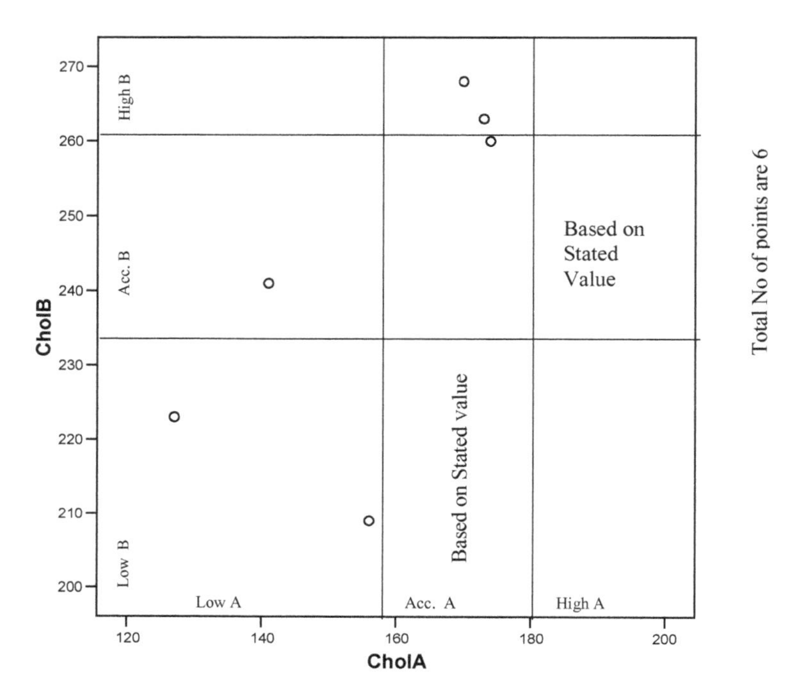

Figure 5.6 Scatter diagram for Total Cholesterol. Plot of sample Total cholesterol A in **Hematrol N** values versus sample Total cholesterol B in **Hematrol P** values.

For the overall quantitative performance of the laboratories, 64.7 % of all of the points on the charts have fallen into the other areas (unacceptable regions). Therefore, it is evident that there was a lack of accuracy as well as precision; and that poor sample handling, poor equipment, poor knowledge of the personnel etc, were responsible for many of the errors.

When we see the overall performance of the participating laboratories based on the above results, by taking individual measurements, *Lab. A, Lab. H and Lab. I* have a better performance in testing the *liver* and *kidney* functional tests as a whole for monitoring HIV/AIDS.

The following scatter diagrams (figure 5.7 – figure 5.12) show the performance of each laboratory on testing the above parameters based on the average values for the three tests of a single parameter in individual laboratories.

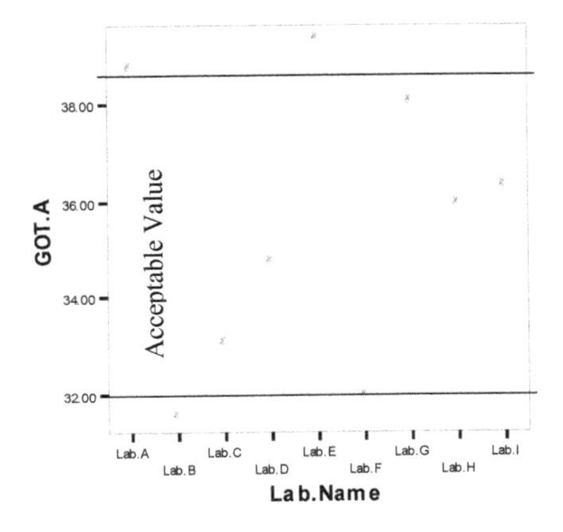

Dot/Lines show Means

From the surveyed laboratories Lab. C, Lab. D Lab. G, Lab. H and Lab. I GOT (A) results fell in the acceptable range and Lab. A, Lab. B, Lab. E and Lab. F GOT (A) results fell out of the acceptable range So, Lab. C, Lab. D, Lab. G, Lab. H and Lab. I had good performance on testing GOT (A).

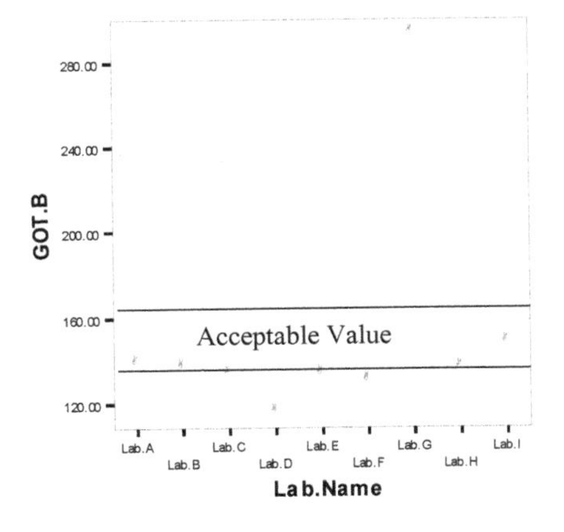

Dot/Lines show Means

For GOT (B) Lab. A, Lab. B. Lab. H and Lab. I had a better performance than the rest laboratories. And GOT (A) was more accurate than GOT (B).

Figure 5.7 The average test results of GOT(A) and GOT(B) for each laboratory

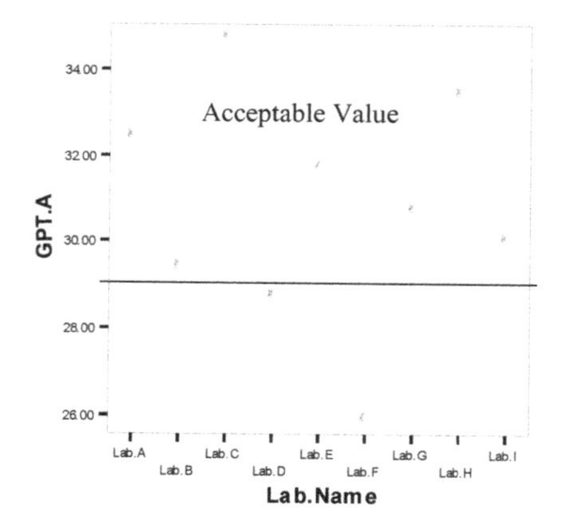

Dot/Lines show Means

For GPT (A) Lab. A, Lab. B, Lab. C, Lab. E, Lab. G, Lab. H and Lab. I had a better performance to test GPT (A) but the rest two Lab. D and Lab. F test results indicated that these laboratories did not accurately determine the test GPT (A).

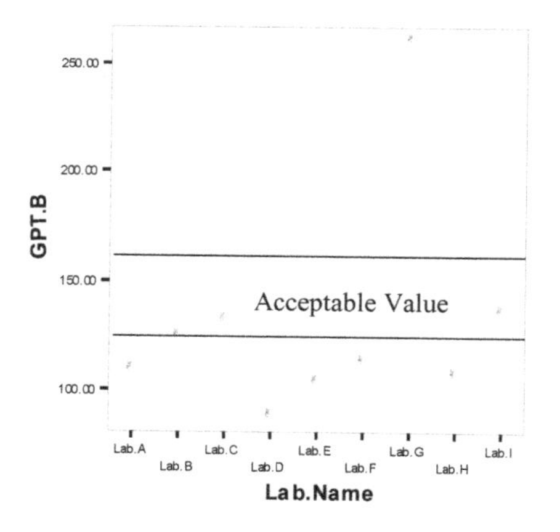

Dot/Lines show Means

The results indicated that only Lab. C and Lab. I have a better performance than Lab. A, Lab. B Lab. D, Lab. E, Lab. F and Lab. H for GPT (B) determination.

Figure 5.8 The average GPT (A) and GPT (B) test results for each laboratory

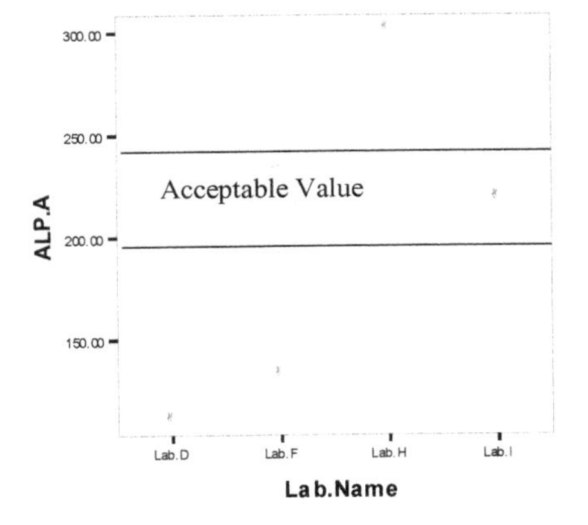

Dot/Lines show Means

For ALP (A) in **Hematrol N** only four laboratories did the test due to lack of reagent. And the results indicated that only Lab. I had a better performance to test ALP (A) but the rest Lab. D, Lab. F and Lab. H test results indicated that these laboratories did not accurately determine the test ALP (A).

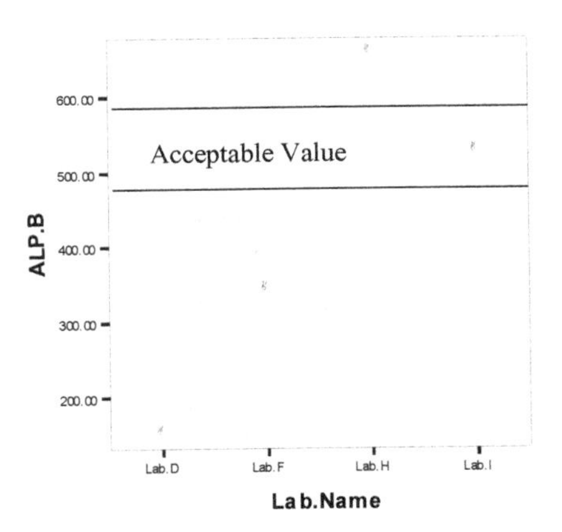

Dot/Lines show Means

For ALP (B) in **Hematrol P** also only four laboratories did the test due to lack of reagent. And the results also indicated that only Lab. I had a better performance to test ALP (A) but the rest Lab. D and Lab. F and Lab. H test results indicated that these laboratories did not accurately determine

Figure 5.9 The average ALP (A) and ALP (B) test results for each laboratory

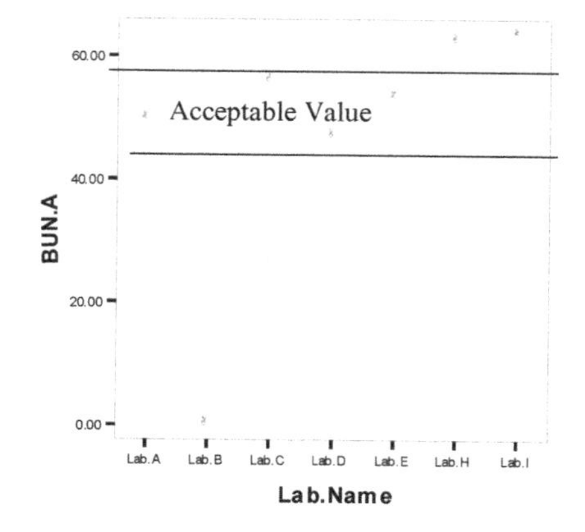

Dot/Lines show Means

For BUN (A) in **Hematrol N** Lab. A, Lab. C, Lab. D and Lab. E, had a better performance to test BUN (A) but the rest Lab. B, Lab. H and even Lab. I test results indicated that these laboratories did not accurately determine the test BUN (A).

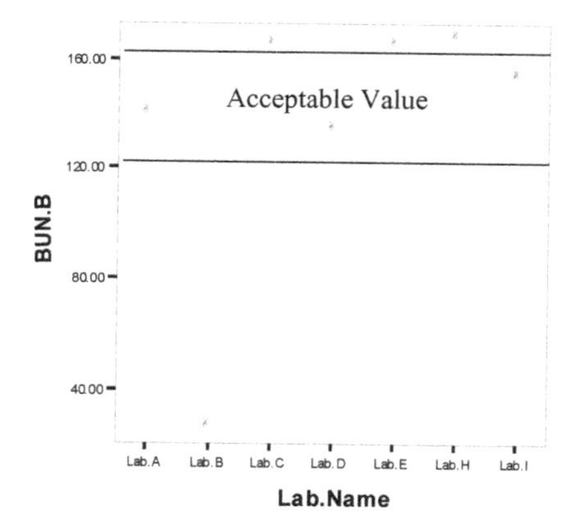

Dot/Lines show Means

For BUN (B) in **Hematrol P** Lab. A, Lab. D and Lab. I had a better performance to test BUN (B).

Figure 5.10 The average BUN (A) and BUN (B) test results for each laboratory

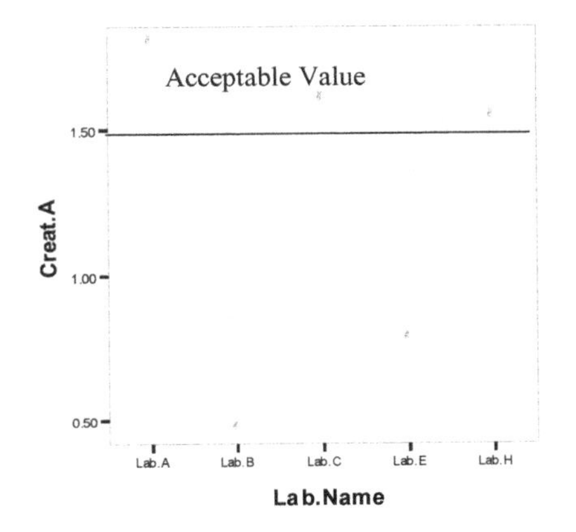

Dot/Lines show Means

For Creatinine (A) in **Hematrol N** Lab. A, Lab. C and Lab. H had a better performance to test Creatinine (A).

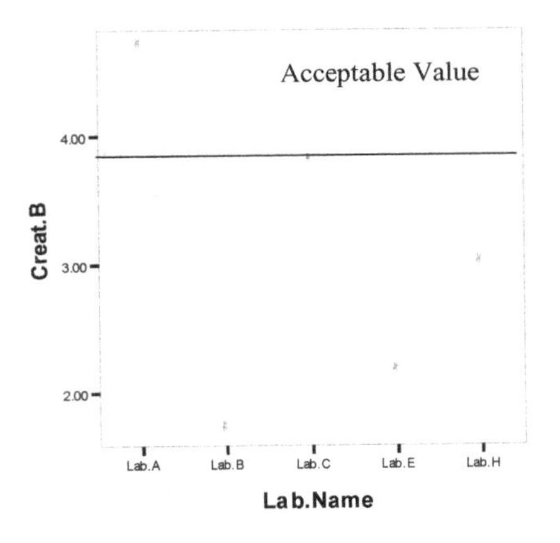

Dot/Lines show Means

For Creatinine (B) in **Hematrol P** only Lab.A had a better performance to test Creatinine (B).

Figure 5.11 The average Creatinine (A) and Creatinine (B) test results for each laboratory

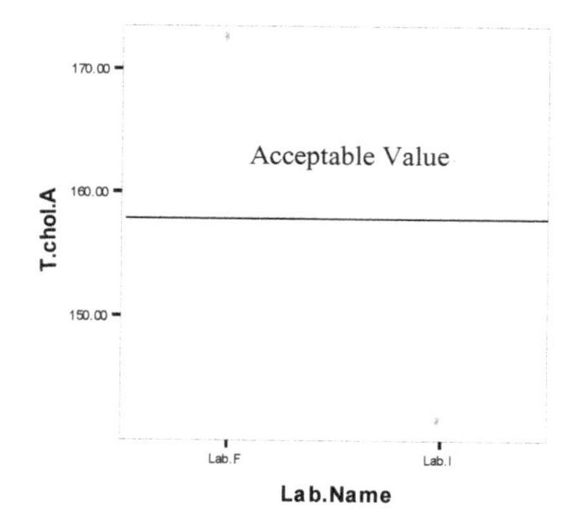

Dot/Lines show Means

For Total Cholesterol (A) in **Hematrol N** only two laboratories, Lab. F and Lab. I did the test. And Lab.F had a better performance than Lab. I to test Total Cholesterol (B).

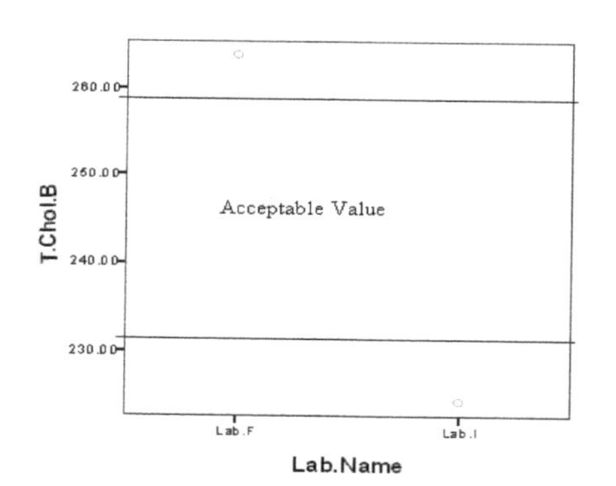

For Total Cholesterol (B) in **Hematrol P** both Lab.F and Lab. I did the test but no one of them results lay on the acceptable region

Figure 5.12 The average Total Cholesterol (A) and Total Cholesterol (B) test results for each laboratory

When we see the overall performance of the participating laboratories based on the above results, *Lab. A, Lab. C and Lab. I* have a better performance in testing the *liver* and *kidney* functional tests as a whole for monitoring HIV/AIDS based on the average measurements and *Lab. A* and *Lab. I* had a good performance both based on *individual measurements* and *average measurements.*

Table 5.1 Summary of the results of all test values in different laboratories

Component	*Sample*	*Stated value*	*No of values reported*	*Mean value*	*Median value*	*Range Of values*	*S.D*	*Allowable limits of error (%)*	*Percentage of Unacce.value (Stated)*	*Percentage of Unacc.value (Mean)*
SGOT/AST	A	35.7	21	35.5	35	27.5 – 48.6	52.5	10	37	37
	B	152	21	152.2	134.6	112 – 330.6	54.9	10	74	74
SGPT/ALT	A	32.2	21	30.8	30.7	22.5 – 41.7	4.3	10	48	48
	B	146	21	143	111.9	85 – 324.2	51.8	10	85	81.4
ALP	A	218	12	220	175.2	88.2 – 322.4	81.7	10	75	75
	B	526	12	510	426.7	152.1 – 683	200	10	75	75
Creatinine	A	1.68	15	1.6	1.56	0.46 – 1.8	0.54	10	53	60
	B	4.55	15	4.8	3.7	1.01 – 4.8	1.31	10	80	80
BUN	A	64.6	18	62	54	0 – 73	20	10	61.8	49.4
	B	147	21	136.1	150	26.1 – 180	48	10	71.4	62
Total – Cholesterol	A	169	6	164	163	127 – 179	19.3	7	50	33
	B	252	6	244	250.5	209 - 268	24	7	66.7	66.7
Average									64.7	61.8

SGOT/AST, SGPT/ALT and ALP values are expressed as IU/L: BUN, Creatinine and Total Cholesterol values are expressed as mg/dl.

The last two columns give the percentage of values classified as unacceptable that is the percentage of the values which fell outside of the ranges calculated from the suggested allowable limits of error. More values fell outside of the ranges based on the stated value than those based on the mean value (64.7 % and 61.8 %).

Table 1 shows that SGOT (A) in **Hematrol N** were estimated most accurately than others and SGPT (B), ALP(B)and Creatinine (B) in **Hematrol P** were the most difficult to determine correctly. The percentage of unacceptable values based on the stated and mean value from 37 – 85 and 37 -81.4 respectively indicates a need for a great deal of improvement in clinical chemistry laboratories to monitor HIV/AIDS properly. In addition to this as one expects more errors were occurred based on the stated value than the mean value and the data also indicated that more errors were appeared on testing sample B than A. so, using **Hematrol N** is advisable for calibration and it is better to take the average of two or more measurements than a single measurement value.

Table 2 shows the general distribution of the results according to their magnitude of error. The data shows that most of the values were grossly inaccurate.

Table 5.2 Classification of all values according to their magnitude of error.

	For errors based on stated value		For errors based on mean value	
Classification of errors	No of values	Percent of values	No of values	Percent Of values
Error < allowable error	101	74.3	91	68
Error > allowable error	35	25. 7	43	32

6. Conclusions and Recommendations

Conclusions

The main purpose of this study was to assess the accuracy and precision of the west Amhara clinical chemistry laboratories on HIV monitoring tests and to determine whether or not there was a definite need for improvement in the performance of these clinical chemistry laboratories. All of the participating laboratories in this assessment scheme completed the survey except some tests did not do due to lack of reagent and improper condition for the specimen. Actually the test results reported by the participating laboratories should have been precise and accurate due to more attention given for the study for fear of showing poor performance. Very likely most of the participants gave special attention to the survey specimens. *However, 64% of the values reported by the laboratories in this study were unacceptable. Thus, there is obviously a definite need for improvement.* Perry ill in the Maryland medical laboratory surveys in the US classified a laboratory as performing satisfactorily only when 10% of the reported values fell out of the allowable limits. The Majority of medical laboratories in west Amhara were out of the standards even in the old GCLP standards.

Some of the laboratories did not have adequate reagents and chemicals to do all the necessary tests and due to this health decisions were made only by one or two laboratory test results and this will be difficult in case of sever problems like HIV.

It is known that the survey has exerted a considerable influence on the thinking of the participant laboratory personnel questioning the accuracy and precision of their analyses and appreciated the need for regular quality control and quality assessment schemes in the work. It is useful for every laboratory to take part in periodic EQA scheme such as this one in order to obtain an unbiased evaluation of its performance.

Recommendations

A good health care system needs complete and standardized medical laboratory services. The Ministry of Health and the regional health bureau should start the implementation of laboratory Quality Management System to reinforce the movement towards quality and standardized medical laboratories. Health care organizations must focus on the continuous improvement of quality rather than focusing on individuals as a source of problems. To implement LQMS, managers should become leaders in advocating continuous improvement, investments in education and training should be made to support continuous improvement, recognition should be provided for all health care workers including medical laboratory technologists, clinical laboratory scientists and an open dialogue should be maintained between customers and suppliers.

In addition the results indicated that sample B's are more difficult to determine than sample A's. So, it is better to use **Hematrol N** for calibration purpose than **Hematrol P** and each laboratory should get chemicals and reagents appropriately, laboratories are better to have a proper temperature maintaining system to treat and diagnosis patients at every time.

References

1. Kassu A and Aseffa A. Pattern of out patient laboratory service consumption in a teaching hospital in Gonder, Ethiopia. East Afr Med J 1996; 73(7): 465.
2. Diane P. Francis, MT (ASCP); K. Michael Peddecord, Louise K. Hofherr, J. Rex Astles; William O. Schalla, MS. Viral load test reports, a description of content from a sample of US laboratories. December 2001; 125: 1546-1554.
3. Constantine NT, Collahan JD and watts DM. Retroviral Testing: Essentials for quality control and laboratory diagnosis, 2nd edition. CRC Press Inc. Boca Raton, USA, 1992; 121: 435 – 451.
4. Daw Mi Mi Sein, M.B., D Path (ygn.). Role of clinical pathology laboratories in patient safety. MJCMP 2003: 193 – 194.
5. David B. Tonks. A study of the accuracy and precision of clinical chemistry determination in 170 Canadian laboratories. 1963; 9(2): 217-233.
6. Ministry of health. Accelerated access to HIV/AIDS prevention, care and treatment in Ethiopia: Road map 2007-2008.
7. Daniel Dejene, Mesfin Addissie and Amha Haile. Assessment of quality of adult ART services in public hospitals of Addis Ababa. 2008; 10-15.
8. Jacoueline Adriane Hearica Droste. Therapeutic drug monitoring of HIV treatment, Bridging laboratory and clinical practices. Nijmegen: 1992: 9-17.
9. World Health Organization. The importance of a laboratory quality. WHO Geneva,1998; 4-9.
10. Jemal Ali, MPH. Assessment of readiness of higher clinics in Addis Ababa, Ethiopia, to initiate ART services. 2008: 1 – 3.
11. Mapunjo S, Urassa DP, et al, Quality standards in provision of facility based HIV care and treatment: a case study from Dares Salaam region, Tanzania. Pubmed (PMID: 17907755).
12. Cheesbrough M. District laboratory practice in tropical countries (part-1). Tropical health technology UK, 1998; 31-37.
13. World Health Organization. Maintenance and repair of laboratory, diagnostic imaging and hospital equipment. WHO, Geneva, 1994; 12-15.
14. World Health Organization. Website http:// www.ac.org.html. Complexity of a laboratory processes.
15. Fuzeon, Summary of product characteristics. Roche Registration Limited, Hertfordshire, United Kingdom: 2003.

16. Riebling N, Tria, Six Sigma Project reduces analytical error in an automated lab. MLO Med Lab Obs. 2005; 37(6):20, 22–23.
17. Daw Mi Mi Sein, M.B., D Path (ygn.). Role of clinical pathology laboratories in patient safety, MJCMP 2003: 193 – 194.
18. Ray Fitzpatrick, Astrid Fletcher, Sheila Gore, David Jones, David Spiegel Halter, David Cox, Quality of life measures in health care. I: Applications and issues in assessment. BMJ 1992; 305: 1074 – 1076.
19. Laffel G, Blumenthal D, J, The case for using industrial quality management science in health care organizations. AM Me Assoc. 1989; 262: 2869-73.
20. Lucia M. Berte., D. Joe Boone, Ph. D, Greg cooper. CLS, MHA, Daniel W. Tholem, M.S, A quality management system model for heath care; approved guideline, 2nd edition. 24, 1 – 49.
21. Berwick DM. N, Continuous improvement as an ideal in health care. Engl J Med 1989; 320: 53-56.
22. O'Connor SJ, Service quality: understanding and implementing the concept in the clinical laboratory. Clin. Lab. Manage Rev: 1989; 3: 329- 335.
23. WHO. Introduction. Overview of the quality system. Module 1. Content sheet 1 -2: overview of the quality management system.
24. 42 CFR, Part 493, Subpart K (493.1252), Standard: Test Systems, equipment, instruments, reagents, materials, and supplies. Oct. 2005.
25. 42CFR. Part 493, Subpart K (493.1239), Standard: General laboratory systems quality assessment. Oct. 2005.
26. Kubono K, Outline of the revision of ISO15189 and accreditation of medical laboratory for specified health checkup (in Japanese). Rinsho Byori. 2007; 55: 1029 – 1036.
27. Clement E. Zeh, Seth C. Inzaule, Valentine O. Magero, Timothy K. Thomas, Kayla F. Laserson, Clyde E. Hart, Jhon N, Field experience in implementing ISO 15189 in Kisums, Kenya. Nkengasong, and the KEMRI/CDC HIV Research Laboratory, 2010; 134:410– 418.
28. Ethiopian health and nutrition research institute. A five year, balanced score card based strategic plan (2010 – 2015 G.C): 10 – 19.
29. Downer K, Five years after "To err is human" clinical lab news. 2005; 31: 1-7.
30. Centers for disease control and prevention (CDC), centers for medicare and Medicaid services (CMS), HHS, medicare, Medicaid, and CLIA programs, laboratory requirements relating to

quality systems and certain personnel qualifications. Final rule, Fed Regist. 2003; 68: 3640-3714.

31. Baltimore, Centers for medicare and Medicaid Services, CMS state operation manual: Interpretive guidelines. 2004
32. James O. Westgard, and Stem A. Westgard, Ms, The quality of laboratory testing today. An assessment of σ-metrics for analytic quality using proficiency testing surveys and the CLIA criteria for acceptable performance. Am J Clin Pathol 2006; 125:343-354.
33. Belete Tegbaru, Hailu Meless et. Al. The status of HIV screening laboratories in Ethiopia. Original article.
34. Phillip L, Internal quality control and external quality assurance in the IVF laboratory. Matson. 1998; 13: 156 – 160.
35. 29 CFR. Part 1910, Subpart Z (1910.1030), Blood borne pathogens. 2005 http://www.access.gpo.gov/nara/cfr/waisidx_05/29cfr1910a_05.html.
36. http://www. Acas. Org/ treatment. Monitoring tests. Common blood tests for monitoring your health. December 2005, page 1 – 3.
37. Diane P. Francis, MPH, MT (ASCP); K. Michael Peddecord, Louise K. Hofherr; J. Rex Astles; William O. Schalla, MS, Viral load test reports, a description of content from a sample of US laboratories. 2001; 125: 1546-1554.
38. World Health Organization. Guidelines for organizing National External Quality Assessment Schemes for HIV serological testing, WHO, Geneva, September 1990.
39. Hoffman RG. Establishing quality control and normal ranges in the clinical laboratory, 1st edition, exposition press, New York, 1971;1-102.
40. World health organization (2006). The urgency of pain control in adults with HIV/AIDS: HIV/AIDS cancer pain release.
41. Hogg RS, Health KV, Yip B, Craib KJ, O'Shaughenessy MV, Schechter MT, Montaner JS, Improved survival among HIV infected individuals following initiation of antiretroviral therapy. JAMA 1998; 279: 450-454.
42. World health organization, The urgency of pain control in adults with HIV/AIDS: HIV/AIDS cancer pain release. 2006.
43. Website http:// www. Cdc.gov/ idu. html. Basic factors about HIV/AIDS.
44. Farr, J. Michael, Laurence Shatkin, Best jobs for the 21st century. JIST Works, ISBN 1563709619; 2004: 460

45. St Clair MH, Richards CA, Spector T, Weinhold KJ, Miller WH, Langlois AJ, Furman PA, 3'- Azido- 3'- deoxythymidine triphosphate as an inhibitor and substrate of purified human immunodeficiency virus reverse transcriptase. Antimicrob Agents Chemother. 1987; 31: 1972-77

Appendix

Appendix A: Laboratories with their corresponding representation

Debark--- Lab. A

Debremarkos --- Lab. B

Debretabore--- Lab. C

Feleghiwot--- Lab. D

Funetesela -- Lab. E

Gondar --- Lab. F

Metema/Shadie -------------------------------------- Lab. G

Mota---Lab. H

Bahir Dar Laboratory Research Center------------Lab. I

Appendix B: Clinical Chemistry lab. Tests result slip used for data gathering.

Bahir Dar University

Science Collage

Chemistry Department

P.O Box 79 Tel 0911781774

Bahir Dar

Clinical Chemistry Lab. Tests Result Slip

Lab ID. No---------------------

Date: Dispatch---------------

Reported---------------------

Name of Equipment---------------------------------------Type of Method---------------------------

Specimen------------------------------

No	Tests	Result	Normal Range	Remark
1	SGOT/AST			
2	SGPT/ALT			
3	ALP			
4	Urea			
5	Creatinine			
6	Total Cholesterol			

Performed By: ------------------------------------ Sign------------------ Date-------------------

Appendix C

Methods used to determine ALT or AST, BUN and Creatinine

1. Determination of ALT or GPT and AST or GOT International federation of clinical chemistry (IFCC)

Preferable specimen: - serum or plasma.

Method of determination: - Auto lab and Human star 80

Reagent

R_1 Tris buffer pH 7.8-----------100 mmol/l

L-Aspartate-------------------200 mmol/l

LDH----------------------------800 mmol/l

MDH -------------------------600 mmol/l

R_2 NADH--------------------------0.18 mmol/l

2-Oxoglutarate----------------12 mmol/l

Procedure

- L-Asparte + 2-Oxoglutarate $\xrightarrow{AST/GOT}$ Oxalacetate + L-Glutamate
- Oxalacetate + NADH + H^+ $\xrightarrow{MDH}$ L-Malate + NAD^+

Prepare the working solution as follows:

1ml of **R_2** to one vial of **R_1** (10 ml vials)

2 ml of **R_2** to one vial of **R_1** (20 ml vials)

5 ml of **R_2** to one vial of **R_1** (50 ml vials)

- Pipette into test tube or cuvette Working solution 1000 µl and Serum or plasma 100 µl.
- After 1 minute run the solution in Auto lab or Human Star 80

2. Determination of blood urea nitrogen (BUN)

(Bertholet method)

Preferable Specimens: - serum, plasma or urine.

Method of determination: - colorimetric method or uv- enzymatic method.

Reagent: - Enzyme reagent, color reagent, base reagent, Standard, 25 mg/l

Materials Required

1. Accurate pipetting devices
2. Timer
3. Cuvettes
4. Spectrophotometer
5. 37°C heating bath

Procedure

Transfer 0.5 ml of *color reagent* to vials labeled: *unknown*, *control*, *standard reagent*, and *black*.

2. Add 0.010 ml of sample to its corresponding vial.
3. Add 0.5 ml of *enzyme reagent* to all vials, mix gently, and incubate at 37°C for five minutes.

(Alternative: React for 10 minutes at room temperature 2- 26°C).

4. Add 2.0 ml of *base reagent*, mix and incubate at 37°C for 5 minutes.

   (Alternative: React for 10 minutes at room temperature 2-26°C).

5. Set the wavelength of the photometer at 630 nm. Read and record the absorbance of all vials.

Calculations

$$\frac{A(U)}{A(S)} \times C(S)\ mg/dl = C(U)\ mg/dl$$

Where: A = Absorbance, U = Unknown, S = Standard, C = Concentration:

Example: A (U) = 0.31, A(S) = 0.48, C(S) = 25 mg/dl

Then:

$$\frac{0.31}{0.48} \times 25\ mg/dl = 16\ mg/dl$$

3. Determination of creatinine (Jaffe reaction)

Preferable Specimen: - plasma, serum, urine, whole blood, and other body fluids.

Method of determination: - Auto labs, Human star 80.

Reagent: - Creatinine R1 Reagent: Alkaline Buffer, Creatinine R2 Reagent: Picric Acid 40mM, Surfactant

Materials required

- Micropipette measuring
- Centrifuge for serum separation
- Test tube and test tube rack
- Refrigerator
- spectrophotometer

$$\text{Creatinine + Sodium Picrate} \xrightarrow{\text{Alkali}} \text{Creatinine-picrate complex (yellow-orange).}$$

1. Prepare working reagent (combine five volumes of R_1 and one volume of R_2 reagent)
2. Set the spectrophotometer cuvette temperature to 37°C.
3. Pipette 1.0 ml of working reagent into each tube.
4. Zero spectrophotometer with the reagent blank at 510 nm.
5. Add .05 ml of sample to reagent, mix and immediately place into cuvette.
6. After exactly sixty seconds read and record the absorbance (A_1)
7. At exactly sixty seconds after the A_1 reading, again read and record the absorbance (A_2), i.e. the time elapsed between A_1 and A_2 is sixty seconds.
8. Calculate the change in absorbance (Δ Abs/min) by subtracting (A_2-A_1).

CPSIA information can be obtained
at www.ICGtesting.com
Printed in the USA
LVHW040500230623
750589LV00005BA/117